The Colour Science of Dyes and Pigments

The Colour Science of Dyes and Pigments

K McLaren

Second Edition

Adam Hilger Ltd, Bristol and Boston

British Library Cataloguing in Publication Data

McLaren, K.
The colour science of dyes and pigments
1. Colour
I. Title
535.6 QC495
ISBN 0-85274-490-0

First published 1983
Second edition 1986

Consultant Editor: **Professor W D Wright**

Published by Adam Hilger Ltd
Techno House, Redcliffe Way, Bristol BS1 6NX, England
PO Box 230, Accord, MA 02018, USA

Printed in Great Britain by J W Arrowsmith Ltd, Bristol

Contents

Foreword

The number of books on colour which are being published each year is a clear indication that the subject is very much alive and that it is being studied in ever-increasing breadth and depth. Certainly no one can read this book by Keith McLaren without being struck by the vitality with which he treats his subject and by the breadth and depth of his own understanding of it. He takes us from Stone Age man to microprocessor robot, and en route deals with the physics and chemistry of dyes and pigments, the visual processes by which their colour is perceived, the instrumental and computer techniques available to industry for the measurement of colour and the prediction of the results of colorant mixture, and finally with those knotty problems that have been with us for so long, the specification of colour differences and the measurement of colour appearance.

In all these fields Mr McLaren writes with the authority born of experience and I cannot think of anyone better qualified to cover the subject so effectively. After extensive industrial experience in the Dyestuffs Division of Imperial Chemical Industries, he moved to Instrumental Colour Systems where he has been very actively involved in the development of colour measuring instruments and their efficient use by industry. He has been a leading figure in the Society of Dyers and Colourists and his contributions to its activities have been recognised by the award of the gold medal of the Society and the research medal of the Worshipful Company of Dyers. In further recognition of his 'outstanding contributions to the study of colour fastness and colour measurement of textiles' the Society also elected him an Honorary Fellow in 1970. Internationally, he has played a very important part in the work of the Colorimetry Committee of the Commission Internationale de l'Eclairage, especially in the discussions leading to CIE recommendations on uniform colour spaces and colour difference equations. At the time of writing this Foreword, he is just completing his two-year term of office as Chairman of the Colour Group (Great Britain).

Now through this book his influence will extend to a still wider audience of colour scientists and industrial colourists, to the great benefit of us all. We have much to thank him for.

W D Wright

Preface to Second Edition

Developments in colour science and technology have continued at a rapid rate since the typescript of the first edition was completed. The span of these developments ranges from the elucidation of the molecular structure of the visual pigment, rhodopsin, to the adoption of instrumental shade sorting by the garment industry of the UK. It was therefore thought more appropriate to issue a second edition rather than a second printing, especially as it permitted the inclusion of many items which, for one reason or another, were omitted from the first.

K McLaren

Preface to First Edition

Colour has fascinated Man since the dawn of civilisation as it was people of the Old Stone Age who painted lifelike figures of animals in many shades of red, yellow, brown and black on cave walls. During the Bronze Age the art of dyeing was discovered and since then pigments and dyes have provided most of the colours of the man-made world. This world of colour involves many different branches of science.

It is, of course, the *chemical* structure of the colorant which governs its colour which appears as a result of the interaction of light and matter – *photophysics*. Such interactions are also capable of destroying not only the colorant (fading) but also textile fibres (tendering). This is the realm of *photochemistry*. Light absorbed by any one of the four pigments in the retina of the eye results first in a change in cell potential followed by a signal being sent to the visual cortex in the brain – *neurophysiology*. How the brain interprets these neural signals brings us into the realm of *psychology* and the fusion of this with physics, *psychophysics*, has provided us with a method of measuring the million or more colours the eye and brain can perceive. The science of colour measurement has also made it possible to predict how much of which colorants would have to be mixed together to match any one of these colours to an accuracy which at least equals that of the professional colourist.

This book gives an overview of all of these scientific and technological aspects of colour, but does not attempt to give a comprehensive account of any. The number of references which have not been mentioned are many times greater than the number which have.

Many individuals have, directly or indirectly, helped in the writing of this book and I would particularly express my thanks to four of them: to the late Deane B Judd whose many publications on colour widened my horizons beyond those of a colour chemist; to W D Wright without whose encouragement this book would never have been started and whose detailed comments after reading the original manuscript resulted in many improvements; to my chairman, A C Perry who provided me with the opportunity to write it; and to my wife, Patience, who converted my badly written manuscript into an acceptable typescript.

K McLaren

1

A Brief History of the Development of Colouring Matters

1.1 THE DEVELOPMENT OF PIGMENTS

The idea of using naturally occurring materials solely on account of their colour first arose during the Old Stone Age when Neanderthal man dominated Europe between 180 000 and 30 000 BC. Towards the end of this period these peoples began to bury their dead and the excavation of several of their burial sites has revealed that red ochre (anhydrous ferric oxide) had been placed on the cranium and breast of the corpse. After Neanderthal man had become extinct, his European successor, Cro-Magnon man, used much larger amounts of red ochre in burial rites and caves have been discovered where large quantities – several hundred kilograms – had been stored. It is generally accepted that Stone Age man associated the colour of red ochre with that of life-giving blood and the importance of red as a colour is clearly shown by the fact that whenever a language has developed to the stage at which colour-names are devised, the first is always a word for red (Berlin and Kay 1969).

Cro-Magnon man, however, made a much more important discovery: the art of painting as displayed in the famous cave paintings of Lascaux in the Dordogne region of south-western France. Remains of charcoal in the oil lamps used by these painters have been carbon-14 dated at about 15 000 BC. They blended yellow and red iron oxides, soot, black manganese oxide and white porcelain clay to give the desired colours and mixed them with cave water which has a naturally high calcium content ensuring good adhesion and great durability. The paints were then applied to damp, smooth cave walls with brushes made from macerated twig tips and bundles of bison hairs to depict deer, horses, cattle, ibex and bison (Leroi-Gourhan 1982).

For the next ten thousand years, no developments concerning pigments and paints occurred but with the first civilisations of Mesopotamia and Egypt, artists were able to produce a much wider range of colours. The colours most conspicuous by their absence from the Cro-Magnon cave paintings were blues and greens but both became available to Egyptian painters. The most famous blue pigment used by the Egyptians was ultramarine. This was derived from the mineral haüyne (sodium/calcium aluminium silicate) which is mainly responsible for the colour of the gemstone lapis lazuli, both being mined in Afghanistan in 4000 BC. Three green pigments were used – terre verte which was derived from glauconite (potassium/aluminium/iron/magnesium silicate), malachite (copper carbonate/hydroxide) and chrysocolla (copper silicate). Bright reds and yellows were also missing from the pigments used by Stone Age artists but the Egyptians used orpiment (As_2S_3) a bright lemon-yellow mineral and realgar (AsS) a bright red. They were also the first to make pigments which did not occur naturally – white lead (lead carbonate), red lead (Pb_3O_4), lead antimoniate (yellow) and, most important of all, Egyptian blue which was made by heating malachite with sand, chalk and soda.

The following Greek and Roman civilisations, though exploiting to the full the pigments known to the Egyptians, appear to have added only one of their own, the bright red cinnabar (HgS), and it was not until the twelfth century that another pigment was produced. This was a blue pigment, indigo, and in many ways the history of the development of blue pigments is the most fascinating of all and worthy of special consideration.

Blue pigments

We have already seen that natural ultramarine had been in use since 4000 BC and that the Egyptians had introduced the cheaper Egyptian blue. Unfortunately, however, the art of making Egyptian blue had been lost completely since the time of the Romans and was not rediscovered until this century. The expense of ultramarine limited its use to small but important areas in paintings and when the dye indigo was first imported into Europe in the twelfth century, its dark blue colour and its insolubility in water led to it being used for hundreds of years only as a pigment, its major defect being its lack of permanence. In the fifteenth century it was augmented by the manufacture of a vitreous pigment containing cobalt called smalt but this, although permanent, was coarse and lacked intensity. In 1704 a very much better blue pigment appeared which was discovered as the result of a lucky accident. Diesbach, a German paint manufacturer, made his own aluminium hydroxide which he used to form a lake† with the dye cochineal. On one occasion, instead of obtaining a white precipitate, it was blue and the cause was traced to the alum containing an iron salt and the caustic potash containing potassium cyanide which reacted to form complex potassium ferri-ferrocyanide which was marketed as Prussian blue. This is still a major pigment though it is only borderline in respect of its permanence.

Prussian blue was not, therefore, an adequate substitute for ultramarine and in the early nineteenth century the French government offered a prize for the development of a method of producing it synthetically. Just as the discovery of Prussian blue was serendipitous, so was that of synthetic ultramarine. In 1787,

† A lake is a pigment prepared by precipitating a soluble dye as a metallic salt or acid complex, sometimes in the presence of an inert substrate, the latter forming an integral part of the product.

Leblanc developed a process for making soda, one stage being to heat sodium sulphate with coal and limestone. On one occasion a blue encrustation was observed on the walls of the furnace. Following extensive research a method of manufacturing this blue substance was discovered which turned out to be similar in composition to ultramarine whose molecular structure has still not been determined although it is closely related to the zeolites and contains sodium, calcium, aluminium, silica, sulphur and oxygen. Synthetic ultramarine then became the standard blue pigment for artists but it has one major defect, its ease of decomposition by acids which limits its use in paints for protective coatings.

In 1901 another important blue pigment, indanthrone, was discovered accidentally but it will be more convenient to consider its origins in §2.3.

The latest and undoubtedly the last stage in the search for the ideal blue pigment was the result of yet another chance discovery. The dyestuff intermediate phthalimide was made by Scottish Dyes Limited by heating phthalic anhydride and ammonia in an enamel-lined cast-iron autoclave. In 1928, several batches of phthalimide were found to be discoloured and from these a dark blue crystalline substance containing iron was isolated, the iron having come from a minute crack in the lining of the vessel. This compound was remarkably stable and was found to have an entirely new structure, phthalocyanine. When the iron atom was replaced by copper a bright greenish-blue pigment having twice the tinctorial strength of Prussian blue and 20–30 times that of ultramarine was produced. It was marketed as Monastral Blue B by ICI in 1935 and has since become the most widely used blue pigment as its only defect is a tendency to 'bronze', i.e. at certain glancing angles its colour may appear brown.

Pigments of other colours

The development of pigments of colours other than blue followed similar lines. The early pigments usually possessed certain technical limitations, or were too expensive, or were poisonous and they were eventually replaced. Even so quite a number of ancient pigments are still important today, though much improved in quality. Among these pigments are white lead, various iron oxides (ochres) and carbon black. The number of available pigments increased exponentially following the discovery of the first synthetic dye as many of them could be converted into lakes: the lake of Perkin's mauve, for example, was used to print the sixpenny lilac stamp of 1868. Other synthetic colouring matters did not require conversion into lakes. For example, many azo and vat compounds were both dyes and pigments so that today organic pigments outnumber inorganic pigments many times over. Few of those in use during the first half of the nineteenth century were rendered obsolete, however, and they still possess features which have never been imitated by organic pigments.

Until the eighteenth century, the technical properties which an ideal pigment had to possess were those making it suitable for artists who inevitably ground their own pigments in whichever medium was being used. During that century, however, paints became widely used on structures for decorative and/or protective purposes and the pigment's suitability for use in such paints became the most important criterion. Pigments which are far too expensive for domestic paints, such as cobaltous stannate (cerulean blue), are still manufactured for artists' use. Pigments are also extensively used today in printing inks, car finishes, stoving enamels and plastics.

1.2 THE DEVELOPMENT OF NATURAL DYES

After the technique of weaving had been discovered in about 5000 BC it was inevitable that attempts would be made to colour fabrics using the wide range of pigments which were available. These early attempts were not particularly successful because the pigment particles were merely held mechanically between the fibres and were removed by rubbing or flexing and particularly by washing. In the fifteenth century pigments were being permanently applied to fabrics by a technique resembling painting, the pigment being first ground with resin or linseed oil. This technique was quite unsuitable for fabrics in general, however, because the coloured areas were stiff and the paint was prone to cracking if flexed. By the end of the seventeenth century linen dyers had discovered how to form basic copper acetate inside the fibre which produced green shades free from these defects. Later other mineral pigments such as Prussian blue and manganese dioxide (which imparted a bronze colour) were formed *in situ* by textile printing methods.

Bronze Age man, however, discovered a completely new class of colouring matters which could be applied to textile fibres with none of the defects which limited the usefulness of pigments: members of this class are called dyes.

Though which dye was the first to be discovered can only be a matter of speculation, it is highly probable that it was one of those occurring in brightly coloured flowers which could be extracted by water alone to give a coloured solution from which the colour could be transferred to one or more natural fibres directly, i.e. by mere immersion. Such dyes are termed 'substantive' and a list of those which have been identified on ancient textiles is given in table 1.1.

These dyes were, however, very fugitive. If exposed to a few hours of bright sunshine the colour would have faded completely and washing would also

Table 1.1 Ancient dyes possessing substantivity.

Dye	Plant of origin	Fibre	Colour
Anatto	*Bixa orellana*	Animal and vegetable	Orange
Henna	*Lawsonia alba*	Animal	Orange
Safflower	*Carthamus tinctorius*	Animal and vegetable	Red
Saffron	*Crocus sativus*	Animal	Silk – golden yellow Wool – orange

remove all of the dye. Archaeological discoveries at Mohenjo-Daro on the banks of the Indus, however, include pieces of red cloth which have been dated at about 1700 BC and the dye used has been identified as madder whose fastness properties are excellent even by modern standards. Though by no means comparable in importance, the discovery of the madder dyeing method must surely rank with the discovery of bronze itself in ingenuity.

Madder occurs in the roots of *Rubia tinctorum* as a glucoside which is hydrolysed when the dead roots ferment under damp conditions. This process generates alizarin (I). Occurring widely in the valley of the Indus are colourless crystals of alum, and Bronze Age man must have noticed the intense red coloration occurring when alum was added to an extract of fermented *Rubia* roots. Bronze Age dyers then discovered that if cotton fabrics were immersed in a solution of alum followed by immersion in an extract of wood ash (which is alkaline) the fabric could be dyed red by steeping it in a hot solution of alizarin.

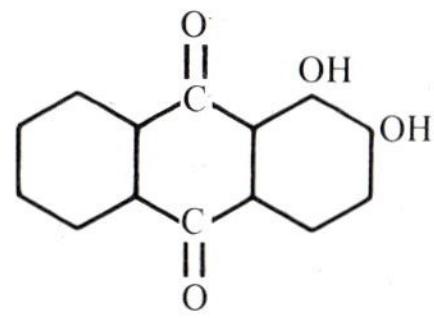

(I)

The effect of the treatment with alum followed by alkali was to deposit aluminium hydroxide in the fibre which in the dyebath combined with the alizarin to form an insoluble red lake. This pre-treatment is called 'mordanting' and its discovery was crucial to the development of natural dyes because so few have any affinity for textile fibres, and most of those which possess some affinity gave much faster shades on mordanted fibres. Alum remained the most important mordant throughout the whole of the natural dye era but salts of other metals were also used to give different shades or better fastness properties: for example, iron and copper salts were in use by the first century AD, tin was introduced in the seventeenth and chromium in the nineteenth century.

Until 1869, when alizarin was synthesised (an event of extreme importance in the development of synthetic dyes), madder remained one of the most important natural dyes for cotton and linen especially when applied by the Turkey red dyeing method. This involved up to twenty separate stages of scouring, mordanting, dyeing and aftertreating which lasted for a total of about a month. Only fibres of the highest quality could withstand such processing and this fact, together with the high fastness obtained, resulted in Turkey red dyed fabrics acquiring the reputation of being virtually indestructible. Greece was the centre of Turkey red dyeing in the seventeenth century and the methods used were not known to the dyers in western Europe. Weavers were therefore forced either to import dyed yarns or to send their yarns to Greece for dyeing. The French discovered the secrets of this method in the middle of the eighteenth century and established Turkey red dyehouses in Rouen and other towns. One stage involved exposing the dyed cloth to sunlight and the area surrounding the site of one such dyehouse at Mulhouse-Dornach is still known as 'La Mer Rouge'.

The Turkey red dyeing process produces in the fibre a complex of alizarin containing both aluminium and calcium, its formula being $CaAl(OH)(Az)_2$ where Az represents the alizarin molecule (I) minus two hydrogen atoms (Kiel and Heertjes 1965).

From the early beginnings in the Bronze Age, dyers discovered many natural dyes from the plant and animal kingdoms – Brunello lists about 300 (Brunello 1973) but in this brief history mention can be made of only those the author considers to be the most interesting, namely indigo, kermes, cochineal, logwood and Tyrian purple.

Indigo

Amongst the most ancient textiles ever discovered are those which have been dyed blue with indigo (II) whose precursor, indoxyl (III), is contained as the glucoside indican in the leaves of two separate species – *Isatis tinctoria* which grows widely in Europe and is commonly called woad and *Indigofera tinctoria* which is indigenous to India. When the dead leaves of these plants ferment, indican is hydrolysed to indoxyl which is readily oxidised by air to indigo. This

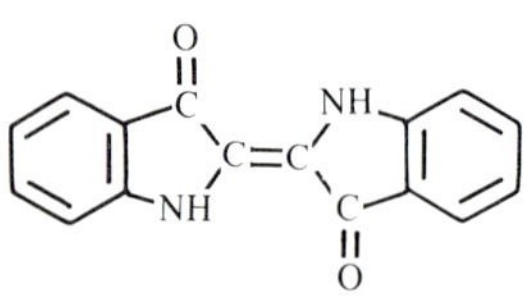

(II)

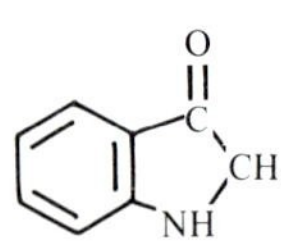

(III)

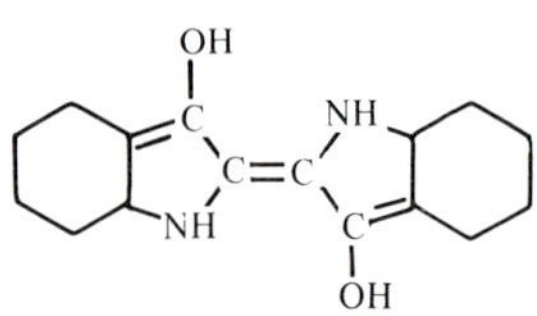

(IV)

technique was used by the Ancient Britons to obtain woad which they used to stain their bodies. Fermentation also causes reduction of any indigo which has been formed to pale yellow leuco-indigo (IV) which is soluble in alkali and possesses substantivity for all natural fibres. Like indoxyl, leuco-indigo is readily oxidised by air to indigo and, after immersion in the dyebath, the fibres are squeezed and allowed to oxidise in air when the characteristic blue colour is formed. The operation of reducing indigo in an alkaline solution was carried out in large vats and this method is termed 'vatting'. Indigo was the only natural dye which required vatting and was one of the few whose fastness properties were not improved by mordanting.

Indigofera leaves contain much more of the glucoside of indican than isatis leaves and in India, where indigo dyeing originated, fermentation was not capable of reducing all the indigo formed by air oxidation. The separated indigo was therefore vatted by a chemical method using arsenic sulphide (As_2S_3) and wood ashes. There are no records to indicate how this unusual method was discovered. One possibility is that as arsenic sulphide was used as a bright lemon-yellow pigment, orpiment (auri-pigmentum), it might therefore have been used with indigo to produce green pigments which might have become alkaline accidentally, the disappearance of the blue component then being observed.

Indigo began to be exported from India to Europe in the twelfth century but as mentioned in §1.1 it was originally used only as a pigment, the first mention of its use as a dye being by Marco Polo at the end of the thirteenth century. It was then being used by Venetian dyers but in virtually every other European country, including England, its use was bitterly opposed by the woad growers and various laws were passed prohibiting its use. This opposition began to crumble when the opening of the sea-route to India in the beginning of the sixteenth century resulted in greatly increased quantities being available at a much lower price. Another factor which undoubtedly helped was that the woad dyers preserved the livelihood of the woad growers, and eliminated the hazards associated with arsenic sulphide, by discovering a better method of vatting. A woad vat was capable of reducing substantial amounts of indigo and this remained a major method for wool dyeing until it was replaced by the use of sodium dithionite in the twentieth century. Cotton dyers, however, abandoned the woad vat in the middle of the eighteenth century in favour of the more readily controlled ferrous sulphate/lime vat whose high alkalinity was not harmful to cotton.

Indigo was probably the most widely used dye during the natural dye era, its importance being indicated by Newton's use of the word as a colour name. The inclusion of indigo as one of the seven named spectral colours has been the subject of much controversy during the present century which has been critically reviewed (McLaren 1985a). The synthesis of indigo in 1880 was another landmark in the development of synthetic dyes.

Kermes

Kermes is another ancient dye having been used to dye the red capes which the Hebrews discovered when they captured the city of Ai in about 1400 BC. It was the most widely used red dye for wool and silk, madder being suitable only for vegetable fibres.

The dye was obtained from the dried bodies of the insect *Coccus ilicis* found on various species of oak growing in the Middle East. The colouring matter present was kermesic acid (V) which was originally applied on an alum mordant. The word *kermes* is of eastern origin and is the root of *crimson* and the French word *carmoisine*. This strongly suggests that the shade it gave on alum was a bluish-red though the colour of the capes discovered at Ai is described in the Bible (*Josh.* VII: 21) as scarlet. Brunello states that in the Middle Ages only the Venetian dyers knew how to produce a scarlet shade from kermes – by mordanting with alum and tartar.

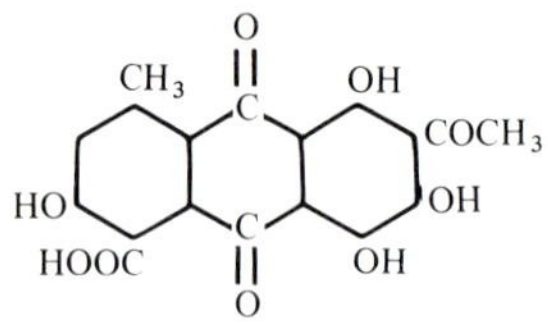

(V)

A dye similar to kermes was discovered by Mexican dyers in about 1000 BC. This was contained in the bodies of the insect *Dactylopius coccus* which lived on the cactus *Rhynchota* and was found being used by the natives when the Spanish entered the country in 1518. The Spanish began to cultivate it and exported it to Europe in 1550 under the name *cochineal* where it rapidly superseded kermes as it contained about ten times as much colouring matter, the glycoside of carminic acid in which the $-COCH_3$ group of kermesic acid has been replaced by $-CO(CHOH)_4CH_3$.

Cochineal certainly gave crimson shades on an alum mordant but in 1607 Drebbel acccidentally discovered that salts of tin gave the much coveted scarlet shade which also had a significantly higher light fastness. Cochineal dyed on a tin mordant resisted the onslaught of synthetic dyes for nearly a hundred years. It was used until 1954 to dye the dress uniforms of the Brigade of Guards though this owed more to innate conservatism rather than technical superiority. Its aluminium lake has been widely used as a food colorant since the Middle Ages and is still widely used for such purposes, being one of the few natural food colorants which has been subjected to toxicological testing. It has never been synthesised on an industrial scale.

Logwood

When the Spanish entered Mexico they discovered that the Mexicans had long been using a dye obtained from a tree growing on the shores of Campeche Bay. The wood from this tree contained the colourless haematoxylin (VI) which oxidised in air to the red dye haematin (VII). When logwood was first exported to Europe, however, there was considerable opposition to its use by the various

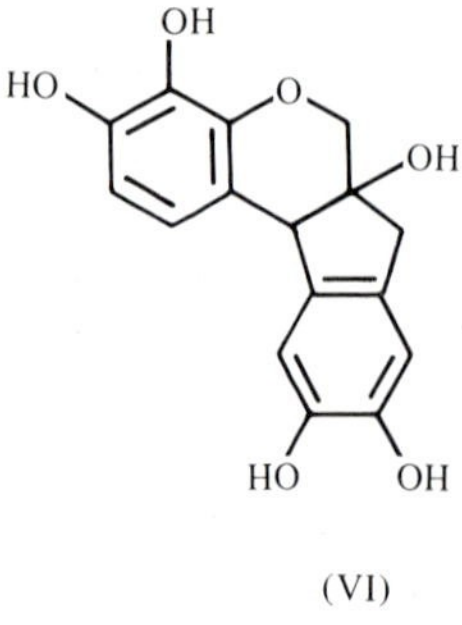

(VI)

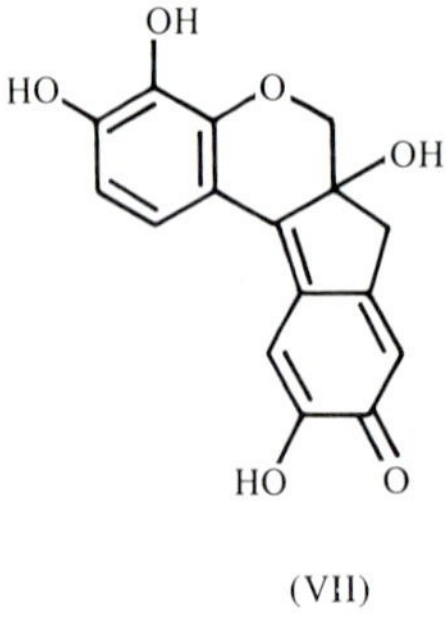

(VII)

dyers' guilds on account of its comparatively poor fastness. In England its use was even forbidden by an Act of Parliament passed in 1580. However, dyers found ways of improving its fastness and by the middle of the nineteenth century it had become one of the most important of all the natural dyes. The shade it gave and its fastness properties depended on the mordant used and it is probable that it was originally used for purple and violet shades on tin or aluminium which are not particularly fast to light. The improvement which occurred during the seventeenth century may well have resulted from its use on an iron mordant which gives dark blue shades of a higher light fastness than any other natural dye. During the latter half of the nineteenth century, chromium became the preferred mordant: this also gives dark blue shades which were often 'topped' with yellow dyes such as *old fustic* to give the bloomy black which still cannot be imitated by synthetic dyes.

Like cochineal, it has never been produced synthetically on an industrial scale but, until recently, it was used industrially as a textile dye.

Tyrian purple

Tyrian purple was discovered in about 1500 BC by the Phoenicians who were the most prestigious of all the dyers of antiquity and during the following three thousand years it has featured significantly in the development and history of several civilisations which flourished on the shores of the Mediterranean – the area comprising Palestine and Syria was called 'Canaan' which means 'land of the purple'. As well as being historically the most important, it is chemically the most interesting of all the dyes ever discovered (Baker 1974).

Several molluscs of the genera *Murex* and *Purpura* secrete pale yellow liquids which contain the precursors of Tyrian purple. According to Pliny the Elder (Bailey 1929) these precursors were extracted from the molluscs by salting them for three days followed by boiling in water for ten days, after which the wool or silk fibres were immersed for five hours. It was then necessary to expose the dyed fibres to light when the colour changed through deep yellow, green, blue to purple. Different species of molluscs gave shades varying from brownish red to deep blue, the most attractive shade being obtained by dyeing initially in a Murex bath followed by a re-dyeing in a Purpura bath. This was the process which gave the 'double-dyed' cloth frequently mentioned in the Bible.

The amount of precursor present was very small about twelve thousand shellfish being needed to obtain two grams which is sufficient for only one metre of cloth. The demand was so great that dyehouses were built all over the Mediterranean and also on the Atlantic coast of North Africa. Tips of waste shells surrounded the dyehouses, some of which were in use for a thousand years. One tip around the important purple-dyeing centre of Tarentum in southern Italy still exists today in the form of the hill called Monte Testaccio composed entirely of *Murex* shells.

Purple-dyed cloth was inevitably extremely costly, the price of a double-dyed purple wool being over a hundred times greater than wool dyed to an inferior bluish-red shade with kermes and indigo. Only the very rich could therefore afford to wear the purple and in Roman times anyone other than royalty wearing a garment dyed with pure purple could be convicted of treason and put to death. With the decline and fall of the Roman Empire purple dyeing virtually ceased in Italy but continued to flourish in the Byzantine Empire, especially around Constantinople. All the earlier dyehouses had to be situated on the coast as only live shellfish were suitable for the purposes of dyeing, but in the fourth century a method of using dead molluscs was discovered which enabled dyehouses to be built far inland where an abundant supply of fresh water existed. One such dyehouse was built in the twelfth century in Upper Egypt. With the fall of Constantinople in 1453 the Byzantine Empire came to an end and the dyeing of Tyrian purple ceased. The basic reaction, however, must have been re-discovered many times subsequently as two instances are recorded in the literature. In 1685 Cole, having heard of someone producing designs on linen garments with 'some liquid substance taken out of a shell-fish' experimented successfully, obtaining a bright crimson shade which withstood more than forty washings (Cole 1685). Nearly two centuries later Lacaze-Duthiers discovered a Minorcan fisherman making designs on white garments using the same shell-fish (Lacaze-Duthiers 1859).

The constitution of the dye present on authentic specimens of cloth dyed with Tyrian purple was established as 6.6′ dibromoindigo (VIII) by Friedländer between 1906 and 1909 (Friedländer 1906, 1909).

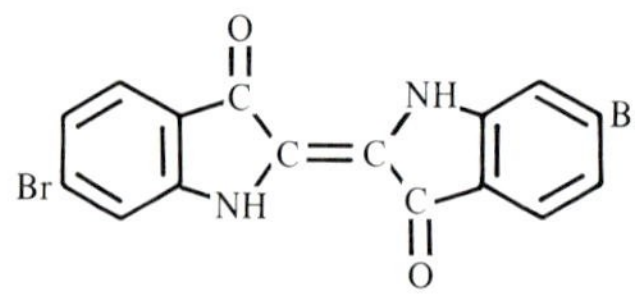

(VIII)

This discovery has resulted in every textbook, and the authoritative Colour Index which will be described in the next section, classifying Tyrian purple as a vat dye with the implication that it was applied to the fibre as leuco-dibromo-indigo.† This is somewhat surprising as none of the many descriptions of the dyeing method by writers such as Pliny the Elder and Pliny the Younger are typical of indigo dyeing because the method of extracting the precursor would inevitably oxidise any leuco-compound. Even more significant is the fact that leuco-indigo is oxidised by air alone whereas exposure to light was an essential stage in dyeing Tyrian purple as confirmed by Cole (1685). None of the hundreds of indigoid dyes which have been synthesised required photodevelopment either, among them being the isomeride of Tyrian purple, 5.5′ dibromo-indigo, first marketed in 1907 as Ciba Indigo 2R and still being manufactured today, its Colour Index number being Vat Blue 35 : 73060.

The molluscs used in dyeing Tyrian purple contain 2-methanesulphonyl 6 bromoindoxyl sulphate (IXa) and the lengthy processes involved in setting the

† In 1984 its Colour Index classification was changed from Natural vat dye: 75800 to Natural Violet 1: 75800.

dyebath cause enzymatic hydrolysis, oxidation and condensation (Fouquet and Bielig 1971) to form *tyriverdin*. This was first identified as a quinhydrone-type complex (Baker and Sutherland 1968, Baker and Duke 1973) but was later shown by Christophersen *et al* (1978) to be the indigoid complex (IXb). This compound yielded dibromoindigo (VIII) on exposure to light at 400 nm with a quantum efficiency high enough to suggest that a chain reaction was occurring (Christophersen *et al* 1978). This reaction also generated a mercaptan which explains Cole's statement that exposure to the sun yielded 'a very strong and foetid smell, as if Garlick and Assa Foetida were mixed together'.

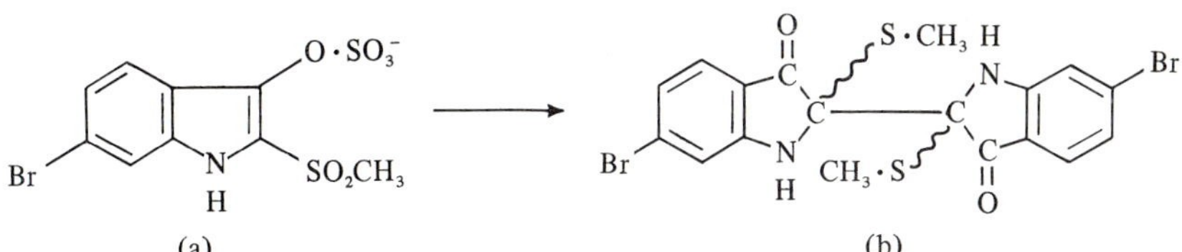

(IX)

Also present in many molluscs is the bromine-free indoxyl sulphate which results in the blue dye indigo (II) and the red isomeride indirubin (XVII) being present on the dyed fibre, varying ratios of the three indigoid dyes accounting for the shades attributed to Tyrian purple ranging from 'brownish red to bluish grey through violet, purple and even deep indigo blue' (Pellew 1919).

Lacaze-Duthiers (1859) used cloth impregnated with Tyrian purple precursor to take photographs and it has been claimed that Alexander the Great used the same photoreaction to ensure that his troops did not suffer from sunburn on the parade ground. His drill sergeants were issued with impregnated arm-bands and were instructed to dismiss the troops as soon as the full purple colour had been developed. This is said to be the origin of the phrase 'Alexander's Rag Time Band' (Beatty 1961).

1.3 THE DEVELOPMENT OF SYNTHETIC DYES

The synthesis and manufacture of a purple dye by W H Perkin in 1856 is rightly regarded as the start of the synthetic dye era. But this was actually neither the first synthetic organic dye to be produced in the laboratory nor even the first to be manufactured. The distinction of being the first synthetic organic dye belongs to picric acid (X), the simplest dye known. It had been vaguely recognised in the eighteenth century as the yellow colouring matter formed by the action of concentrated nitric acid on many organic substances, including skin, and was manufactured in 1845 by nitrating phenol which had been isolated from coal tar. The manufacturer was the senior partner in the Lyon's firm of silk

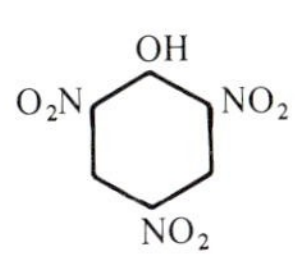

(X)

dyers, Guinon, Marnas and Bonnet and silk dyed with picric acid was exhibited by the firm at the Great Exhibition in London in 1851. Its major use was in combination with blue dyes to give green shades. Although green is the most common plant colour, chlorophyll was not a satisfactory dye and virtually all green shades were obtained from yellow-blue mixtures until 1846 when a homogeneous green dye, lo-kao, began to be imported from China. Lo-kao was a glucoside of unknown composition derived from the shrub *Rhamnus utilis*. No combination of vegetable dyes could match the brilliance of lo-kao because all blues and yellows tended to be somewhat reddish. Picric acid was, however, a bright greenish yellow and by forming Prussian blue within silk fibres dyed with picric acid, Guinon was able to match the shade at a fraction of the cost.

Picric acid is a member of the *acid* dye class which does not occur in nature. The first acid dye, however, was the sodium salt of indigo 5.5′ disulphonic acid (XI) manufactured by Barth in 1740 by treating indigo with concentrated sulphuric acid. It was sold under the name Saxe blue.

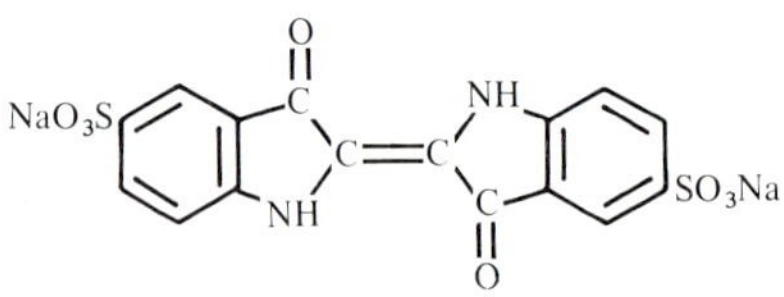

(XI)

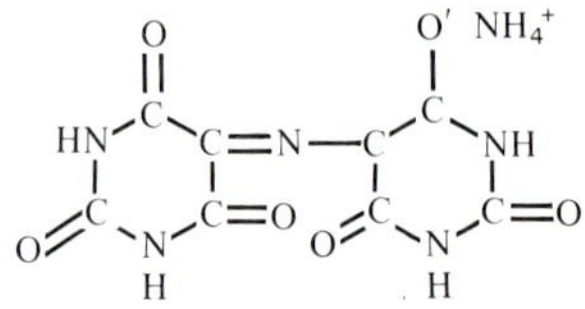

(XII)

The second synthetic dye was murexide (XII) which had been synthesised by Prout in 1818 who noted its potential as a dyestuff. He made it by treating uric acid with nitric acid followed by ammonia but as his source of uric acid was the excreta of a boa constrictor then being displayed in London, its manufacture was not possible until a more abundant source was discovered. Such a source was guano, the dry excreta of sea-birds imported from Peru for use as a fertiliser, and murexide began to be manufactured in 1853 in Mulhouse and in 1855 in Manchester. Its commercial name arose because it was originally believed to be identical with Tyrian purple whose precursor was secreted by *Murex brandaris.* It was applied to tin-mordanted wool, mercury-mordanted silk and lead-mordanted cotton for about twelve years until it was displaced by the easier-to-apply and cheaper dyes derived from aniline, Perkin's starting material.

Aniline purple

Perkin's discovery of Aniline purple was another of those accidental happenings which have so often occurred in the development of colouring matters. Perkin, an eighteen year old assistant to A W Hofmann who occupied the Chair of Chemistry at the Royal College of Chemistry, was attempting to synthesise the anti-malarial drug quinine by oxidising aniline with acidified potassium bichromate. He obtained a copious black precipitate but observed that if this were extracted first with naphtha and then with ethyl alcohol, a purple dye was obtained with affinity for silk. The observation that coloured compounds could be obtained from aniline was not new as Runge had described them over twenty years previously though Perkin was presumably unaware of this. But Perkin was the first to recognise that such a coloured compound might be a dye and to discover that this was indeed the case. This step which sometimes follows a *discovery* justifies the term *invention* though, unfortunately, the difference is often overlooked (Rattee 1965). Together with his father and brother Perkin then established a factory (at Greenford Green in London) for its manufacture. When it is realised that he had to work out processes for making nitrobenzene from benzene (using sulphuric acid and sodium nitrate as strong nitric acid was not available) and for reducing it to aniline which yielded, at most, 2% of aniline purple, the magnitude of his achievement can be appreciated.

Some years later, Heinrich Caro, who we shall later see was one of the great figures of the early dye industry, wrote that 'British inventive genius and enterprise created a new era of scientific and industrial progress. This glorious era of a formerly unknown union of science and industry was inaugurated by Perkin's discovery of the first aniline colour' (Caro 1892).

The structure of the main constituent of aniline purple is believed to be as below (XIII). Aniline purple was a *basic* dye and there was only one natural dye belonging to this class – that derived from the shrub *Berberis vulgaris* which was rarely used. Dyers, therefore, had no established methods of applying this class of dye and Perkin had to develop them and demonstrate them to potential customers, thus beginning the marketing concept of *technical service* which became a standard feature of all the major successors to Perkin and Sons Limited. The dyeing methods he developed were to apply the dye to wool and silk from a soap bath and to mordant cotton and linen with tannin and sodium stannate.

Perkin patented his process (Perkin 1856), the patent being interesting in that 'color' is spelt in the American fashion throughout. The French patent, however, was invalid because of a technical error in filing it. We have seen that one firm of French silk dyers had already been engaged in the manufacture of picric acid. There were, in fact, others: their experience of large-scale processing which

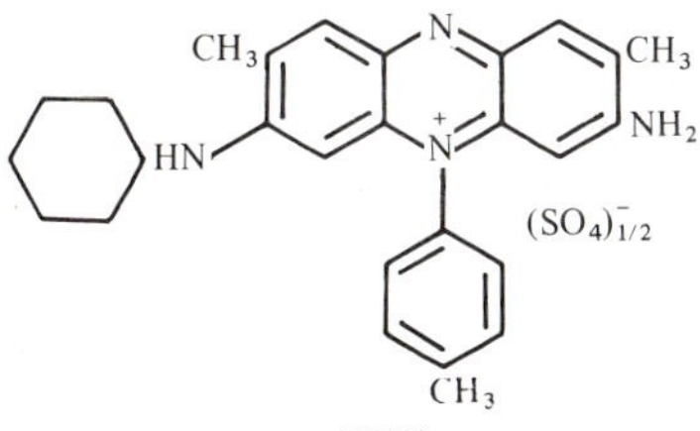

(XIII)

Perkin completely lacked resulted in more than one French firm being able to offer bulk supplies of aniline purple before Perkin and Sons. Poirrier of St Denis named his product after the lilac-coloured mallow flower which the French call 'la mauve'. This appealed strongly to the English world of fashion and mauve became a colour name, the dye becoming better known as 'Perkin's mauve' or 'mauveine'.

Magenta

The fact that a commercially successful dye could be produced from aniline prompted many others, particularly in France, to try the effect of different reagents on aniline. Verguin, who had erected a picric acid plant in 1855, heated impure aniline with stannic chloride and obtained a brilliant bluish-red basic dye which was later named after the town where the French had just achieved victory against the Austrians in the second Italian War – magenta (XIV).

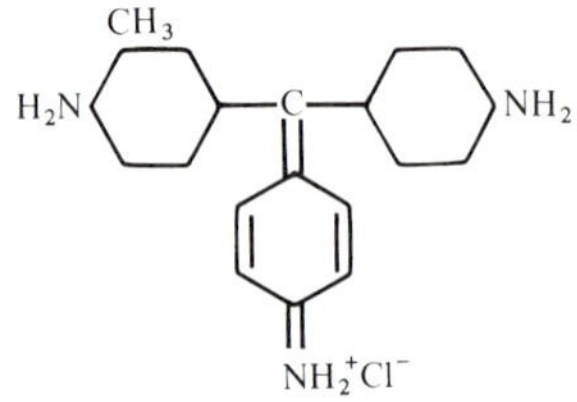

(XIV)

Magenta was much more successful than mauve for two main reasons. The most important was that the shade it gave appealed to the world of fashion more than the duller, bluer mauve, indeed the fashion house of Schiaparelli sponsored it as 'shocking pink'. The second was that the yield of magenta was very much greater than that of mauve. Perkin and Sons immediately recognised its potential and began its manufacture using a method devised by W H Perkin to avoid infringing the French patent.

Magenta was not only a successful dye but was also a valuable intermediate because Girard and de Laire in France phenylated the amino groups to varying degrees to give aniline blue and imperial violet. These were also basic dyes but Nicholson, another graduate from Hofmann's department at the Royal College, applied the sulphonation technique first applied by Barth to indigo, to convert aniline blue into the acid dye – soluble blue. By dissolving magenta in sulphuric acid and treating it with acetaldehyde, an unstable blue dye was formed which Cherpin converted into aldehyde green by treatment with sodium thiosulphate. Dyes were also obtained from other starting materials such as phenol and α-naphthylamine and in 1858 a reaction was discovered which was eventually to yield more dyes than all the others put together. This was the diazo reaction discovered by Griess who became an assistant to Hofmann and eventually chief chemist at Ind Coope's brewery in Burton-on-Trent. The first commercially successful azo dye was Bismarck brown (XV) discovered by Martius whilst working for Roberts, Dale and Company, dye manufacturers of Manchester.

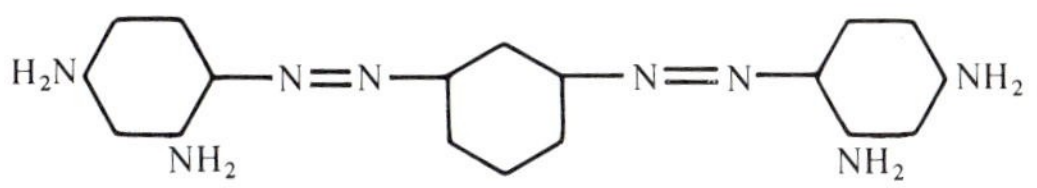

(XV)

It will have been noticed that England and France were the centres of dye development at this time. Germany's contribution consisted of the activities of Hofmann, Martius and Griess whilst in England, together with the exploitation in Germany of English and French discoveries which could not be protected by patents because of the lack of effective patent law before the formation of the Second Reich.

Within thirteen years of Perkin's original discovery it was possible to dye every natural textile fibre in a very much wider range of shades with synthetic dyes, called 'coal tar colours' or 'aniline dyes', than with natural dyes. The synthetic dyes were also much cheaper and easier to apply so it is therefore somewhat surprising to find W H Perkin in his Presidential Address to the British Association in 1876 saying that 'the influence of the so-called aniline colours on dye woods etc has been remarkably small'. The reason for this is not hard to find. All of the dyes discovered up to 1869 were extremely fugitive, a few hours exposure to bright sunshine destroying them completely. They could therefore only be used for the ephemeral world of fashion: for everyday wear and for household textiles, the natural dyes were still essential.

It should not be a matter of surprise that these early synthetic dyes were fugitive because, without exception, they had been produced by methods which were conceptually no different from those used by the alchemists in their attempts to convert base metals into gold. The state of the theory of organic compounds at that time was such that, because the molecular formula of quinine was known to be $C_{20}H_{24}O_2N_2$, Hofmann and Perkin believed that if any compound of this formula were synthesised, the odds were that it would be identical with it! Unlike the alchemists, however, Perkin and Verguin were lucky as their experiments resulted in valuable products being formed which they expoited to the full. It was surprising enough that brilliantly coloured compounds were formed with affinity for fibres – to have expected them to possess good fastness properties as well would have been too much.

An entirely different approach to dye synthesis opened up in 1865, however. This was the theory of the ring structure of benzene proposed by Kekulé who, in spite of his French-sounding name, was a German. This theory had very little impact on the progress of chemical research in either England or France but in Germany its potential value was immediately recognised and within three years it enabled Graebe and Liebermann to establish the structural formula of alizarin, the dye which, when applied by the Turkey red process, was virtually indestructible.

Alizarin

Madder had already been the subject of extensive investigation in the early nineteenth century, the object being to produce it in a more concentrated form. The commercial madder only contained about 4% of alizarin compared with up to 85% of the active agent in commercial indigo. These investigations resulted in alizarin being made in a pure crystalline form in 1831 and identified as a derivative of naphthalene because both products yielded phthalic anhydride on oxidation. It was not, however, a derivative of naphthalene and Graebe and Liebermann correctly identified it as 1:2 dihydroxyanthraquinone (I) in 1868: in the following year they synthesised it. Their method was, however, entirely unsuited for bulk production but within a few months suitable methods were independently discovered by W H Perkin and Heinrich Caro.

Caro was a German industrial chemist who had little academic training but who learnt from his manufacturing experience, first with Roberts, Dale and Company and, from 1868, with one of the first German dye makers, the Badische Anilin und Soda Fabrik (BASF). Amongst his other discoveries were induline, eosine, chrysoidine and methylene blue, which clearly establish him as one of the greatest geniuses in the early dye industry. Caro's British Patent was filed the day before Perkin's but the latter acquired the sole manufacturing rights in the UK and until 1872 Perkin and Sons were actually producing more alizarin than BASF.

The success of synthetic alizarin was immediate because it contained very much more dye than commercial madder and, moreover, commercial madder always contained varying amounts of other polyhydroxyanthraquinones, some of which resulted in dull shades being produced. But the trade in madder was not destroyed overnight because it took several years before there was sufficient manufacturing capacity to meet the world's needs. In 1880, the UK was still importing madder at a rate of about 1000 tons per annum. Perkin and Sons soon realised that their Greenford Green plant could not be expanded to meet the demand but to W H Perkin, the idea of building a new factory had no appeal whatsoever. Perkin and Sons Limited was therefore sold and W H Perkin devoted the rest of his life to organic chemical research. Sir Robert Robinson believes that Perkin was also influenced by his realisation 'that the circumstances in Germany provided the ideal conditions for the most rapid development of this industry, the first not merely to be founded on research but needing to be continually vitalised by research' (Robinson 1956).

The new British factory was subsequently built by The British Alizarin Company at Silvertown, London, in 1874, manufacture being transferred to Trafford Park in Manchester after the 1914–18 war when it became part of ICI. Manufacture ceased there in 1972. The early British and French dye-making industries reached their zenith in 1874. The scientific approach of the Germans and the Swiss together with adequate financial backing led inexorably to Germany and Switzerland replacing England and France as centres of the dye-making industry, the British decline being well illustrated in the following table.

Period	Patents filed by dye-makers	
	British	German
1856-66	141	0
1866-76	59	0
1886-1900	86	948

The British dyestuffs industry, however, was reborn during the First World War (Morton 1929) and is still a world leader.

It should be noted that the synthesis of alizarin was in no way part of the development consequent upon Perkin's discovery of mauve. It was an entirely new point of departure and would have occurred at about the same time even if none of the earlier synthetic dyes had ever been discovered. Around this time another important natural dye, indigo, attracted the attention of chemists.

Indigo

Investigations into the composition of indigo began in the eighteenth century when Woulfe oxidised it with nitric acid and obtained picric acid (X). In 1826, Unverdorben distilled indigo and obtained aniline, whose name is derived from the Sanskrit word for indigo – anil. Between 1870 and 1878 Kekulé made many attempts to synthesise isatin (XVI) which had long been recognised as the com-

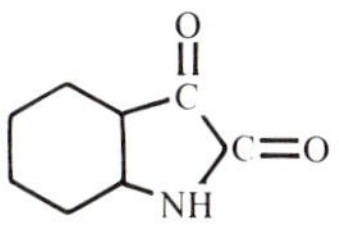

(XVI)

pound obtained when indigo was oxidised under conditions less severe than when using nitric acid. Kekulé was, however, unsuccessful and the challenge was taken up by Adolf von Baeyer who had carried out the experiment of oxidising indigo to isatin in 1848 when he was thirteen years old! Baeyer studied chemistry under Bunsen and Kekulé and came under the influence of Hofmann who left the Royal College of Chemistry for the University of Berlin in 1865. Baeyer succeeded Liebig as professor of chemistry at Giessen where he trained the great majority of the leaders of the German dye industry. He was still teaching up to his death at the age of 80. He succeeded in synthesising indigo in 1878, from o-nitrophenylacetic acid, but this route was completely unsuitable for manufacture. Although he did attempt to discover more practical routes, he was unsuccessful and it was not until 1897 that chemists at BASF established a satisfactory process from phenylglycine-o-carboxylic acid, a venture which has been estimated to have cost nearly as much as the capital value of the company at that time. Although initially priced up to 25% higher than the natural product, it was an immediate success. Like madder, natural indigo contained various amounts of other dyes which affected the shade, particularly the reddish isomeride indirubin (XVII) whereas the synthetic product was much more consistent.

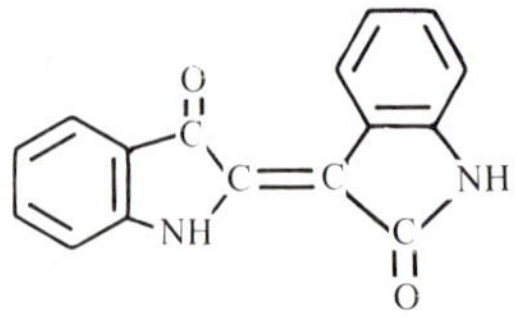

(XVII)

The effect on indigo plantations in India and in North Carolina, which had been established when it was a British colony, was devastating. The production of synthetic indigo or indigotin by BASF in 1900 was equivalent to that from a quarter of a million acres and BASF were not the only manufacturers, another German firm, Meister, Lucius and Brüning having entered the field in 1900 with a more direct route from phenylglycine. The British government afforded some protection by specifying that natural indigo had to be used for naval and military uniforms but the effect was only of minor extent because the indigo used in the UK in 1913 consisted of 208 tons of natural indigo, 300 tons of synthetic indigo made in England and 1194 tons synthetic indigo made in Germany.

Natural indigo was one of the first dyes ever discovered and synthetic indigo which is, of course, chemically identical, is still widely used in spite of the development of many other blue dyes. Indigo therefore spans the whole dyeing era. Recently its main use has been for dyeing cotton yarn to be woven as weft with undyed cotton yarn as warp into denim and much indigo was exported to China for dyeing the familiar 'boiler suit' national dress. In the early 1950s, however, Mao Tse-tung banned all imports of indigo and about 25% of the market vanished overnight. Several manufacturers ceased production leaving only four plants which could easily meet the requirements of the rest of the world. These plants are in Germany, USA, Japan and England. The English indigo factory was built in 1908 at Ellesmere Port on the estuary of the Mersey by Meister, Lucius and Brüning. It was requisitioned by the government in 1914 and eventually became part of ICI. In the mid-1960s denim became fashionable as 'blue jeans' and although there were severe shortages of indigo, no new plants were built. The existing plants are over 50 years old and have been written off against tax many times over. This means that no new plant could possibly produce indigo as cheaply and, in any case, the fickle nature of the world of fashion could suddenly result once again in surplus manufacturing capacity.

The structural formulae of most of the important natural dyes have been determined and confirmed by synthesis in the laboratory. Only one other textile dye had a structure which was amenable to synthesis on an industrial scale – Tyrian purple – though it was the 5.5′ isomeride which was marketed (in 1907) as this was easier to make. Its method of application was of course as a conventional indigoid dye and bore no resemblance to the ancient photochemical dyeing method.

Another naturally occurring colouring matter has been synthesised on an industrial scale – β-carotene (XVIII). This product, and several other carotenoid pigments, continued to be used as food colorants, e.g. saffron, annatto and paprika, in spite of the availability of synthetic dyes. As various natural extracts containing these pigments had been used for centuries, complete freedom from any adverse effects could be assumed, unlike many of the synthetic dyes whose usage was severely restricted or even completely banned in many countries. Synthetic carotenoids were therefore potentially of commercial importance and in 1950, β-carotene was synthesised from acetone in the laboratory (Inhoffen 1950). It was marketed in 1954, an industrial route having been discovered (Isler *et al* 1956) and its commercial success has led to several other synthetic carotenoids being marketed by the Swiss firm, Hoffmann la Roche.

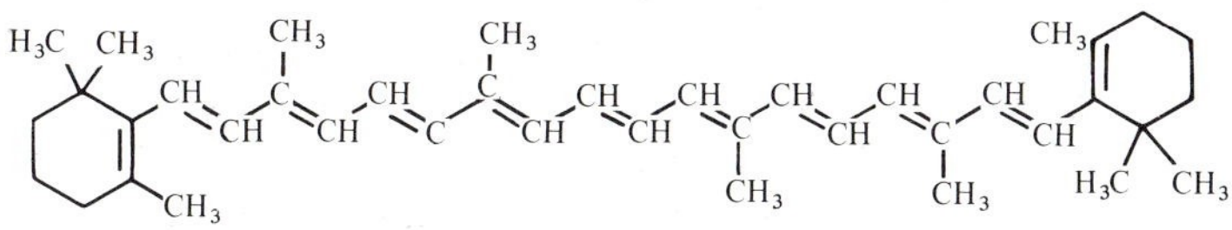

(XVIII)

The names of colouring matters: the Colour Index

All of the natural and synthetic colouring matters so far mentioned have had individual names such as Tyrian purple, magenta, etc but towards the end of the nineteenth century, new dyes were invariably introduced under a brand name – a registered trade mark. When patent cover lapsed, other manufacturers would sell an essentially identical product under a different brand name.

In 1924, the British Society of Dyers and Colourists† published an index of all the colouring matters under the title *Colour Index* a somewhat misleading title as it was a colouring matter or *colorant* index. Second and third editions have been issued jointly with the American Association of Textile Chemists and Colorists‡, the current edition being up-dated by Additions and Amendments issued quarterly (SDC/AATCC 1971). A history of the *Colour Index* and a prediction concerning its future has recently been published (Burdett 1982).

Every colouring matter, natural and synthetic, is given a generic name which usually indicates the application class, e.g. Pigment Red, Acid Yellow, Vat Blue, followed by a serial number. Each entry contains fastness data and the index lists every brand name under which it is marketed. If the chemical constitution has been disclosed, this is given under a five-figure constitution number together with references to publications including patents. From this point onward, every colorant mentioned will be identified by its *Colour Index* name and, if appropriate, its number.

Manufacturer's brand names, commonly called commercial names, invariably give information of interest to the user. For example, Vat Yellow 1 was first

†Perkin House, Bradford BD1 2JB, England.
‡PO Box 12215, Research Triangle Park, NC 27709, USA.

marketed as Indanthren Yellow G, Indanthren being the brand name given by BASF to anthraquinonoid vat dyes of very good all-round fastness properties, the letter G indicating that it gave shades which were greener than reddish-yellow dyes which would be designated R, 2R etc or as Golden Yellows. Bird has published a detailed account of the large amount of data which have been incorporated into brand names (Bird 1945).

The growth of synthetic dyes

The development of synthetic dyes can best be appreciated if the relationship between the structural formula of a dye and its suitability for various end-uses is understood. These topics will be discussed in chapter 3. This chapter will be concluded by some statistics concerning the growth of the synthetic dye industry.

The total number of dyes and pigments which have been marketed is about 10 000 and, of course, this number is only a small fraction of the number synthesised in laboratories but not worthy of manufacture. Holbro has said that the Swiss firm, Ciba, had synthesised well over 10 000 azo pigments of a particular type but that only about twelve were ever marketed (Holbro 1962) giving a ratio of 1 in 1000.

The azo class is pre-eminently the one which lends itself to the synthesis of large numbers of speculative dyes but even if the overall ratio is 1 in 400 it would still mean that dye chemists have synthesised about four million dyes. In 1965, when the total number of dyes marketed was about 6000, Rattee, when he held the Chair of Colour Chemistry at Leeds, estimated the speculative total at three million (Rattee 1965). The number of dyes synthesised must surely be many times greater than all the other synthetic organic compounds put together.

Natural dyes had to be applied singly. If it were necessary to use two dyes, for example when dyeing green shades before the introduction of lo-kao, two quite separate dyeing operations had to be carried out. It might therefore be thought that the very large number of dyes marketed was partly due to the needs of fashion. This, however, is not the case. Synthetic dyes of the same dyeing class can be used together in the same dyebath and fewer than ten dyes, if used singly and in two- or three-dye combinations, will produce the complete gamut of colours possible from the whole range. The reasons for the large number of dyes will be discussed in chapter 3.

In 1960, White of ICI Dyestuffs Division estimated that in the whole world there were about 1000 chemists seeking new dyes and he seriously doubted if the 'unsatisfied needs of the textile industry (could) sustain such an army of speculative dye chemists' (White 1960). The annual average number of new dyes which were marketed between 1850 and 1982 is shown in figure 1.1 most of the data being obtained from the various editions of the *Colour Index* (McLaren 1982). The peak of 1974 was due to a sudden influx of Japanese dyes which had been discovered over many years but not submitted for entry into the

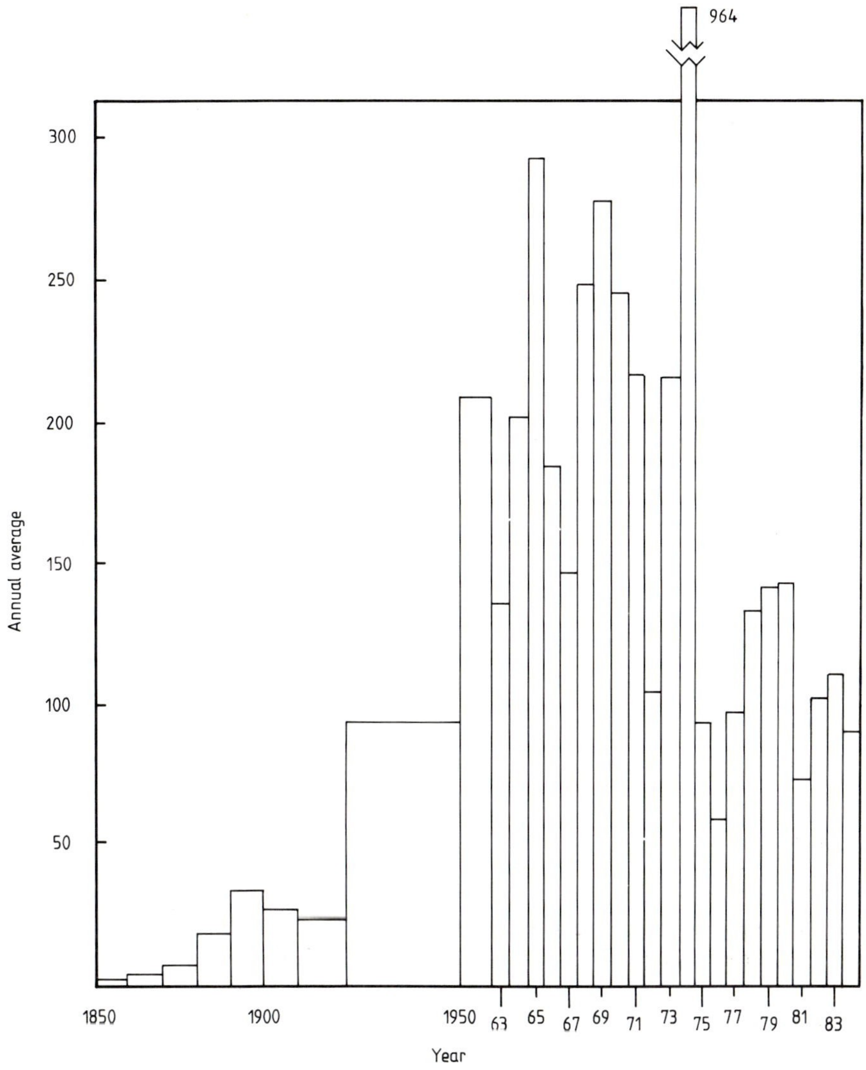

Figure 1.1 New dyes marketed between 1950 and 1984.

Colour Index until the Japanese government required this as a pre-requisite for sale in Japan. Figure 1.1 shows that White's doubts began to be confirmed about ten years later when other factors, particularly the cost of ecological testing, began to have an effect. Unfortunately there does not appear to be any chance that the growth rate curve will do other than approach zero before the end of the century. Since 1974, seven major American chemical companies have withdrawn from the dye-making industry (Sullivan 1982).

2

The Photophysics and Photochemistry of Colouring Matters

2.1 COLOUR AND CHEMICAL CONSTITUTION

The first theory relating colour and chemical constitution was put forward in 1868 by Graebe and Liebermann. This stated that as all known dyes were decolorised on reduction, colour was associated with unsaturation. Witt (1876) extended this by postulating that all coloured organic compounds must contain a group which introduced a potentiality for colour. These groups he termed *chromophores* and as examples he cited $-NO_2$, $-NO$, $-N{=}N-$, $>C{=}O$ and $>C{=}C<$. This potentiality was then developed by weak salt-forming groups which he termed *auxochromes,* e.g. $-OH$, $-NH_2$, $-NHR$ and $-NR_2$, which also conferred dyeing properties.

In spite of its extreme simplicity, Witt's theory dominated virtually the whole of the synthetic dye era. As soon as a new chromophore was discovered, literally thousands of compounds containing it and various auxochromes and other substituents previously found to be of value, were synthesised and evaluated, the promising ones being manufactured. This approach has been aptly described as 'enlightened empiricism'. Fewer than 25 chromophores have been identified, the most recent being that responsible for the colour of the important phthalocyanine blue pigment mentioned in chapter 1. This provides a perfect example of the difference between a discovery and an invention. Copper phthalocyanine was first discovered in 1927 (Diesbach and Weid 1927) but they did not attempt to elucidate its structure or its properties. Iron phthalocyanine was then discovered independently as described in chapter 1 (Dandridge *et al* 1929) but they found that iron, copper and nickel phthalocyanines were potentially valuable pigments – an invention.

The structure of phthalocyanine was elucidated by a series of brilliant investigations performed by Linstead and his colleagues (Dent *et al* 1934, Linstead and Lowe 1934) and confirmed by one of the earliest applications of x-ray analysis (Robertson 1935). The structural formula of the metal-free phthalocyanine molecule (XIX) is shown below. It contains a strainless sixteen-membered central ring exhibiting marked planarity.

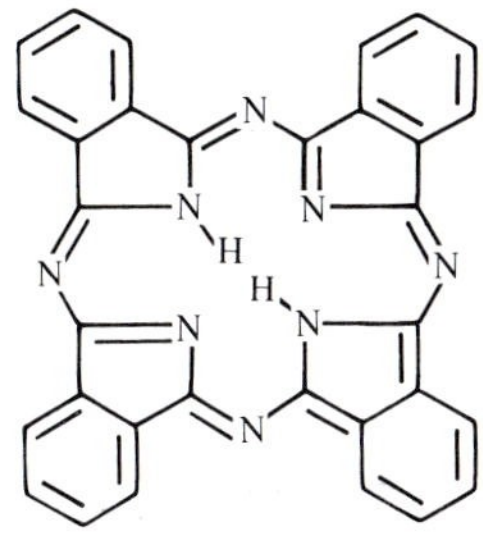

(XIX)

Phthalocyanine readily forms co-ordination complexes with metals giving CI Pigment Blue 15:74160 (XX) and two naturally occurring pigments – haem (XXI), responsible for the red colour of blood and chlorophyll (XXII) the green plant pigment (CI Natural Green 3:75810).

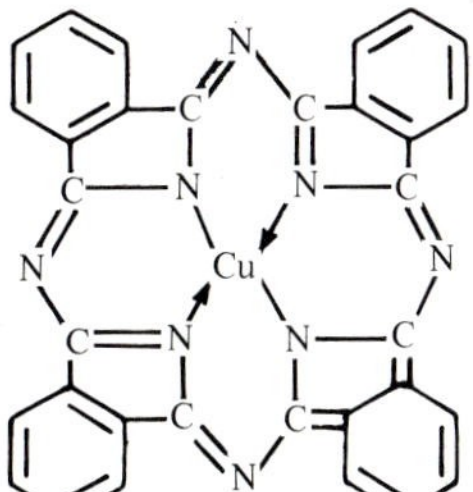

(XX)

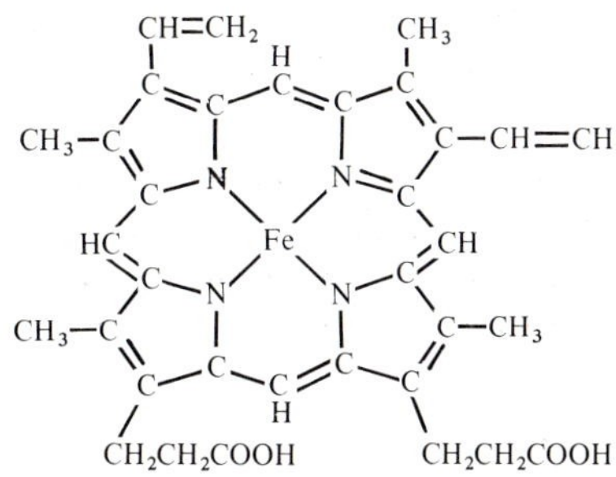

(XXI)

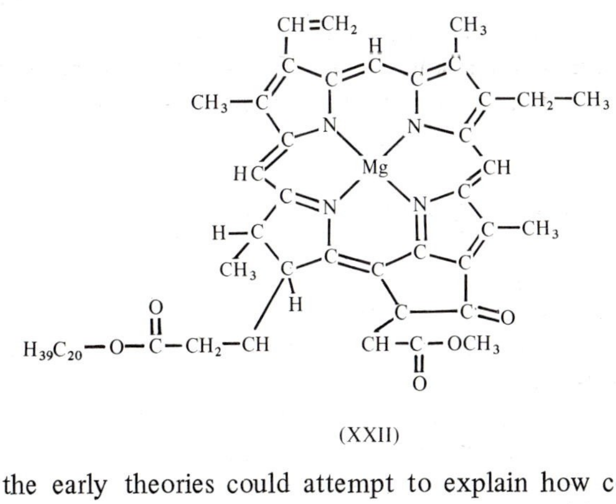

(XXII)

None of the early theories could attempt to explain how colour arose in organic compounds. An explanation only became possible when it was discovered that electromagnetic radiation in the ultraviolet and visible regions interacted with organic compounds by a photon being absorbed whose energy promoted an electron into an orbital of higher energy forming an excited state. All coloured organic compounds contain carbon–carbon double bonds and these are formed by the overlapping of two sp^2 hybrid orbitals of each carbon atom to form an ordinary carbon–carbon bond called a σ bond and a second chemical bond called a π bond. These bonds are molecular orbitals, each being occupied by two electrons which, in accordance with the Pauli exclusion principle, have opposite spins. At the same time as these σ- and π-bonding orbitals are created, two other anti-bonding orbitals are also formed designated σ^* and π^* which are unoccupied when the molecule is in the dark, in its ground state. Another orbital may be involved in colour formation if the molecule contains a nitrogen or an oxygen atom in certain positions – this is the orbital occupied by two electrons which are not involved in bond formation, i.e. 'lone pair' or n (non-bonding) electrons. Each of these orbitals has a specific energy level and these are represented simply by an energy level diagram.

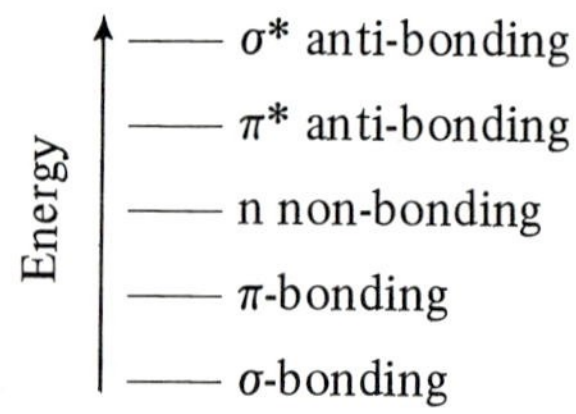

Each electronic level has a number of other levels associated with it and these will be described later.

When ultraviolet or visible radiation is absorbed, an n, σ or π electron is promoted to occupy an empty π^* or σ^* anti-bonding orbital, the wavelength, λ, of the photon being such that its associated energy is equal to the energy difference, ΔE, between the two levels involved namely

$$\lambda = hc/\Delta E$$

where h is Planck's constant and c is the velocity of light. Transitions to a σ^* orbital can be ignored as the energy required corresponds to radiation in the far ultraviolet but a π or an n electron can be promoted to a π^* anti-bonding orbital by visible light. Three such transitions are important as far as colorants are concerned: n-π^*, π-π^* and C–T (charge transfer). A C–T transition occurs when a conjugated system contains an electron-donor substituent such as $-NH_2$ or $-OH$ and an electron-acceptor substituent such as $-C{=}O$ or $-NO_2$. In such cases electron promotion is accompanied by an intra-molecular charge transfer causing a marked increase in the polarity of the excited state and a marked change in chemical reactivity.

Whenever the energy difference is equal to the energy of visible light, the compound has an absorption band in the visible region and has the colour complementary to that of the visible light absorbed. This simple statement, without elaboration, occurs in many textbooks but it is totally inadequate, as the colour of a dye or pigment also depends on factors other than the wavelength of the absorption band.

Factors affecting colour

The molar extinction coefficient

The fundamental method of measuring the intensity of the colour of a compound is to dissolve it in a suitable solvent and to measure the fraction of light transmitted (the transmittance) at the wavelength of the absorption peak. From this fraction, the molar extinction coefficient, ϵ, can be calculated from

$$\epsilon = \lg \text{transmittance}/cl$$

where c is the concentration in moles per litre and l the length of the optical path in centimetres. The molar extinction coefficient is actually a measure of the probability that a photon of the correct wavelength striking the molecule will be absorbed: if the probability is 100% the molar extinction coefficient is about 10^5. The values for most dyes are about 10^4 and these are among the highest found in any type of substance. The human eye can readily see the colour of an amount as small as 10^{-18} g or about 10 000 molecules.

Obviously the higher the molar extinction coefficient the better, but a high value is by no means essential. If dye A has a lower molar extinction coefficient than dye B but costs less to make or has greater affinity, other factors being equal, A may be the better choice for marketing, because the lower coefficient can be overcome by using more of it. Dye manufacturers do not, therefore, go to the trouble of making routine determinations of extinction coefficients, though a comparison of the density of a solution against a standard is widely used.

The wavelength of maximum absorption, λ_{max}

When a single absorption band moves from the ultraviolet into the visible spectrum the compound has the colour which is the complementary colour of that of the light absorbed as shown in table 2.1.

Table 2.1 Light absorbed and colour of absorbing compound.

Light absorbed		Colour of compound	
Wavelength (nm)	Colour	Qualitative	Metric hue angle†
400–440	Violet	Greenish yellow	115
400–500	Blue	Yellow	90
460–500	Greenish blue	Orange	65
400–620	Bluish green	Red	20
480–520	Green	Magenta	360
560–700	Orange	Cyan	215
600–700	Red	Bluish green	195

† The quantification of hue by its metric hue angle will be described in chapter 11.

The early dye chemists regarded the progression from greenish yellow through red to bluish green as 'deepening' but this is misleading as an intense yellow is obviously a deeper colour than a pale turquoise. Such a shift is better described as 'bathochromic' and one in the reverse direction as 'hypsochromic', these terms being first proposed in 1892.

In spectroscopy it is common practice to describe a bathochromic shift as a 'red shift' and a hypsochromic shift as a 'blue shift' but when dealing with colorants these terms should normally be avoided because it seems illogical to describe a shift from the violet region of the spectrum to the blue as a red shift, for example.

It is obvious from table 2.1 that certain important colours cannot be obtained from a single absorption band, the most important being green. This requires the presence of two bands, one in the red and one in the blue-violet. This is much more difficult to achieve than single band absorption and in consequence the number of homogeneous green dyes and pigments is significantly less than that of any other major colour (figure 2.1). Black dyes are also extremely important and they require a combination of several broad overlapping bands covering the whole of the visible spectrum, the extinction coefficients of the various bands being similar. Brown, olive green and other dull colours also require bands covering the whole visible spectrum but in such cases different extinction coefficients are essential. It is interesting to note that few organic compounds absorb strongly in the far red and hence strong solutions of most dyes appear red.

The width of the absorption band

The width of the absorption band is of much greater importance than the extinction coefficient and this is well illustrated by glass containing the rare earths neodymium and praseodymium. This is widely used to calibrate the wavelength scale of spectrophotometers under the name 'didymium glass', a name which is a reminder that at one time didymium was thought to be an element. This glass has a very strong absorption band in the yellow region whose peak is at 580 nm but it is only pale blue in colour because the band is narrow, being 30–40 nm wide at half the height of the peak. Glass containing cobalt has an absorption peak at about the same wavelength but because it is about 200 nm wide at halfpeak, the colour is an intense blue. The absorption bands in table 2.1 are of such a width as to give colours of the maximum intensity.

The profile of the absorption band

The profile of the absorption band in the visible region is extremely important as it governs the *chroma* of the perceived colour. Chroma is one of the three fundamental variables of perceived colour which will be described in chapter 7 but for the present it may be more readily recognised by related terms such as saturation, intensity or purity. The maximum chroma possible would be given an absorption band(s) whose profile is vertical, and whilst many dyes in

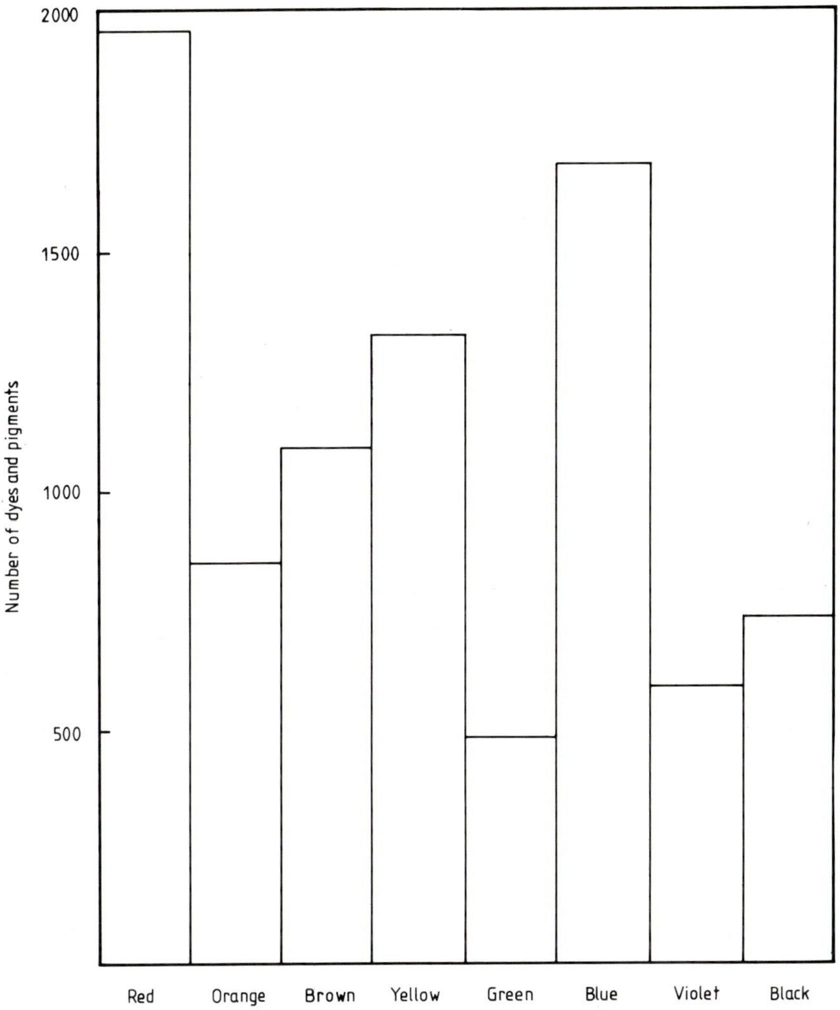

Figure 2.1 The number of dyes and pigments of each major colour (from *Colour Index* vol. 5 1982).

solution do have near vertical profiles, when applied to a substrate the profiles become much less steep with the sole exception of yellows. This seriously limits the attainable chroma by causing unwanted absorption. For example, every non-fluorescent red dye reduces significantly the amount of red light reflected by the undyed textile.

The commercial implications of this are very serious in the two industries (colour printing of non-textile materials and colour photography) which are based on the subtractive primary colours, yellow, magenta and cyan. All known magenta and cyan dyes absorb surprisingly strongly in the spectral regions where ideally no absorption should occur. High quality colour printing therefore requires the use of more than three colours whilst the best negative-positive photographic print processes employ a very sophisticated technique known as 'integral masking' to overcome the effects of unwanted absorption (Hunt 1967).

It is not known why yellow colorants should be comparatively free from this defect. The science of light absorption by organic compounds has made it possible to predict only the wavelength of the main absorption peak in isolated cases, but not peak width or profile. In view of the millions of dyes synthesised, however, it is safe to assume that the cause lies in the complexities of electronic excitation, a subject which is far from being fully understood.

The spectral refractive index

Refractive index is of importance only in the case of pigments as it governs the extent to which incident light is scattered. With dyes all of the scattering is caused by the substrate. The effectiveness of a white pigment is governed by the extent to which its refractive index (RI) exceeds that of the medium in which it is dispersed. Calcium carbonate has an RI of 1.6 and is effective when applied from an aqueous dispersion – whitewash – as the RI of air is 1.0. The RI of a typical paint medium, however, is also about 1.6 and when calcium carbonate is dispersed in such a medium it contributes very little to the opacity and its use is that of an extender or filler. White lead, at one time the traditional white paint pigment, has an RI of 2.0 and the technical superiority of titanium dioxide, first introduced in about 1916, is due to its higher RI, 2.7. In complete contrast, a pigment can only be effective as a black if its RI is close to that of the medium: that of aniline black (CI Pigment Black 1 : 50440) for example is 1.4.

The refractive index of a white or black pigment is virtually independent of wavelength, but if a pigment has an absorption band in the visible region this results in the RI varying according to the wavelength of light used for determination – hence the term spectral refractive index. This is a fundamental characteristic dependent primarily on chemical constitution and as it cannot be altered for a given pigment or be easily determined it has received little attention. There seems to have been only one systematic investigation – that of Cooper who determined spectral RI for sixteen pigments (Cooper 1948). He found that when RI was plotted against wavelength, the curve shapes were often quite different for pigments of similar colours. For example, Prussian blue had a low RI over most of the spectrum with a rapid rise at the extreme red; phthalocyanine blue, on the other hand, had a pronounced V-shaped curve. Though there was little correlation between the shapes of RI and absorption curves, the wavelength of any pronounced peak did indicate the hue perceived when a surface coating exhibited 'bronzing', the metallic lustre that is seen at near-glancing angles. This is usually reddish-yellow in colour as many RI curves peak in this region: phthalocyanine blue, however, having two peaks at the ends of the spectrum, exhibits a purplish 'bronze'.

Examples of electronic excitation

Lycopene

Lycopene is the pigment responsible for the colour of ripe tomatoes. It has the simplest molecular structure (XXIII) of all the millions of organic colorants because it consists of carbon and hydrogen atoms only and contains no ring structure. Its absorption spectrum has a strong, broad band at 480 nm, $\epsilon = 18 \times 10^4$. Witt's theory fails to explain the colour of lycopene because it does not contain any group Witt would have considered to be a chromophore or an auxochrome. The correct explanation has important implications outside this comparatively small group of colorants.

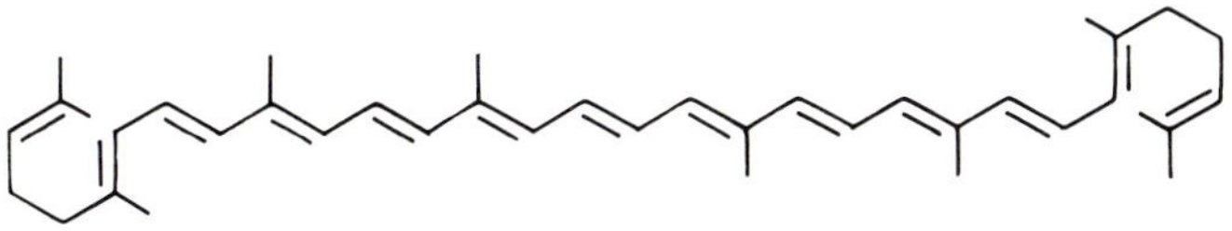

(XXIII)

If a paraffin chain contains one carbon-carbon double bond, e.g.

$$CH_3{-}CH_2{-}CH_2{-}CH{=}CH{-}CH_3$$

its absorption spectrum contains a strong band at about 165 nm, $\epsilon = 10^4$, caused by a π-π^* transition. The presence of a second double bond, e.g.

$$CH_2{=}CH{-}CH_2{-}CH_2{=}CH{-}CH_3$$

does not result in a second band appearing because the energy involved in each of the two π-π^* transitions is exactly the same: the spectroscopist describes such a situation as 'degenerate'. The intensity of the band is increased somewhat however, as the chance of photon capture is obviously greater. If the two doubly bonded carbon atoms are separated by a carbon-carbon single bond as in 1.3 butadiene however

$$CH_2{=}CH{-}CH{=}CH_2$$

the situation is now entirely different. The π orbitals not only overlap the doubly bonded carbon atoms, they overlap the singly bonded atoms as well, reducing the distance between them from what it is in a paraffin chain. The effect of this is that the energy required to cause a π-π^* transition is reduced ($\lambda_{max} = 209$ nm, $\epsilon = 2.5 \times 10^4$) and the presence of additional doubly bonded carbon atoms which preserve the alternation of single and double bonds reduces the energy required still further. There is in consequence a bathochromic shift, and when the number of doubly bonded carbon atoms reaches six, the absorption peak extends into the visible region. A chain of alternating singly and doubly bonded carbon atoms is termed 'conjugated', the classic case of course being that of the benzene ring. The effect of increasing the length of a conjugated chain on λ_{max} is illustrated by the series of β-apo-carotenals (XXIV) as λ_{max} increases steadily from 413 nm when there are seven carbon atoms in the chain to 508 nm when the number is thirteen.

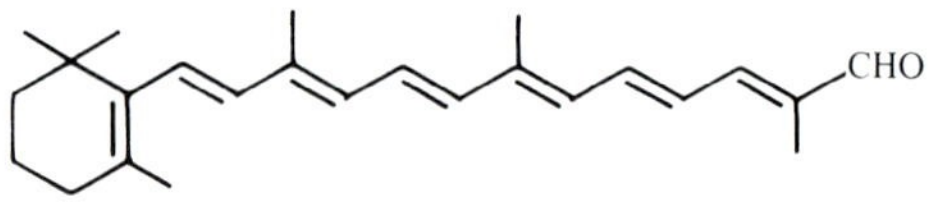

(XXIV)

The azo group

For about 80 years the azo group was thought to provide the strongest evidence in support of Witt's theory and the reasons for this are not hard to find. Every known azo compound is coloured: even the simplest, diazomethane (XXV) is a yellow gas. The simple azo compounds are not strongly coloured in either the original sense of the term, i.e., the absorption band is in the violet region, or the preferred sense, the extinction coefficient is low. The presence of an auxochrome, however, intensifies the colour in both senses of the term.

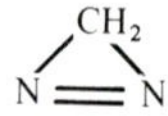

(XXV)

The azo group contains two pairs of non-bonding electrons and one electron can be displaced into a π^* anti-bonding orbital by the absorption of visible light. In the case of azo-benzene (XXVI) λ_{max} is at 440 nm but ϵ is only 5×10^2.

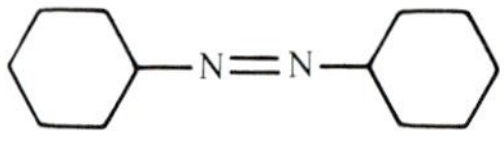

(XXVI)

Introduction of an auxochrome greatly intensifies the strength of the absorption band in the visible – but it is now known that it is not the n-π^* band localised on the azo group which has been intensified. Figure 2.2 shows the absorption spectrum of azo-benzene in methanol. The two broad, strong, overlapping bands are due to π-π^* transitions of the delocalised electrons of the whole molecule and it is these bands which undergo a bathochromic shift when one or more auxochromes are present. This band then completely swamps the weak n-π^* band and in no commercial colorant is it believed that such a band makes any significant contribution to the colour.

The azo group is of inestimable value in dye synthesis mainly because it can link two conjugated systems of benzene or naphthalene rings to form a single conjugated whole. The effective length of the conjugated whole can be readily extended by the introduction of two or more azo groups. The *Colour Index* lists

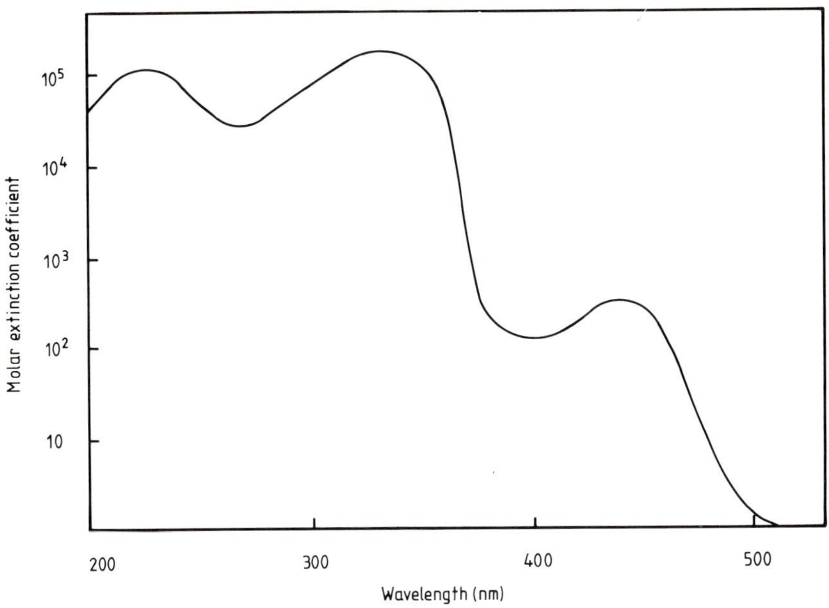

Figure 2.2 Absorption spectrum of azo-benzene in methanol.

fourteen dyes containing five or more azo groups, the record being held by two brown cotton dyes with eight each. The multiplicity of benzenoid rings gives ample scope for the introduction of auxochromes for either modifying the colour or, as Witt envisaged, conferring dyeing properties though these two effects are not necessarily related as we shall see in chapter 3.

The mode of action of auxochromes

To understand the mechanism by which an auxochrome causes a bathochromic shift it is necessary to regard the conjugated system as capable of existing in two or more electronic structures, i.e. the resonance forms. None of these structures describes the molecule as it really is but it may be considered as a hybrid to which the imaginary structures contribute. An auxochrome functions by increasing the stability of one or more of the resonance forms and it does so because it can retain an electrical charge more readily than a carbon atom can. This increasing stability permits the π electrons to move more readily along the chromophoric group and this reduces the energy required for excitation. Groups which reduce the stability have a hypsochromic effect.

A chromophoric block

In complete contrast to the effect of the azo group in conjugation preservation, an ingenious use for a group which did not link two chromophoric systems into a single whole was discovered by the Swiss firm Ciba. The group was trichlorotriazine (XXVII) which readily combines with dyes containing amino groups by

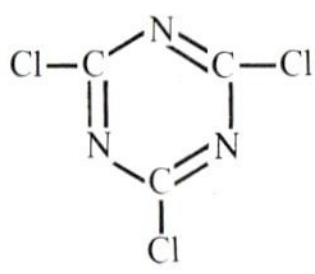

(XXVII)

a Schotten-Baumann reaction to give dyes of the general formula (XXVIII). It is now impossible to pass from Dye 1 to Dye 2 via atoms which are linked alternately by single and double bonds: the two dyes therefore preserve their individual conjugations, the triazine ring acting as a 'chromophoric block'. By using it to link yellow and blue dyes together, valuable green dyes were discovered which are still marketed, for example, CI Direct Green 28:14155.

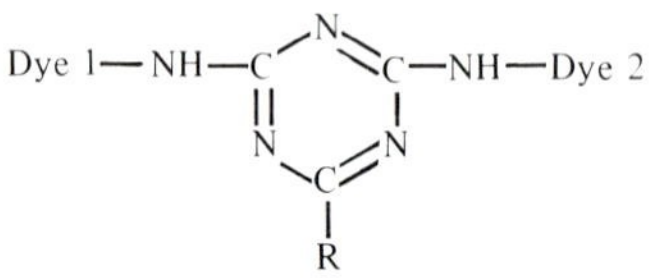

(XXVIII)

Recent developments

The advent of the quantum theory led to major developments in the field of the relationship between colour and chemical constitution and several major theories have been proposed. Space does not permit an adequate review to be given here and the interested reader is referred to the following authors (Coates 1967, Mason 1970, Griffiths 1976, Dahne and Kulpe 1978, Klessinger 1978, Fabian and Hartmann 1980, Griffiths 1984).

2.2 PHOTOPHYSICAL PROCESSES FOLLOWING LIGHT ABSORPTION

When an electron has been displaced into a higher orbital, the molecule becomes highly unstable because of the energy absorbed. This is surprisingly large – a molecule which has absorbed a photon of ultraviolet light at 300 nm, the short wavelength limit of daylight at sea level, has absorbed the same amount of energy as if it had been heated to about 1500 °C. It must therefore rid itself of this excess energy and it can do this in a number of different ways, both physical and chemical. In this section we shall consider the photophysical processes involved but before doing so it is necessary to consider in greater detail the nature of the electronic states of molecules.

The solution of the wave equation for the hydrogen atom as formulated by Schrödinger gave rise to three parameters which, with a fourth arising from Dirac's relativistic mechanics, are called quantum numbers. Three of these define an electronic orbital and the fourth, the spin of an electron in its orbital which can have two values, $-\frac{1}{2}$ and $+\frac{1}{2}$. The Pauli exclusion principle states that two electrons cannot occupy the same orbital if they have the same values for these four quantum numbers. The n and π orbitals of colouring matters contain two electrons and these must therefore have opposite spins. The resultant spin angular momentum is zero giving rise to a single energy level and each electronic state in which all the electrons are paired in spin is termed a *singlet*, represented by $S_{0,1,2,3\ldots}$, according to the energy level.

Associated with each electronic energy level are a number of higher levels which are defined by vibrational quantum numbers and between each of these are a number of quantised rotational energy levels which are conventionally represented in the following manner:

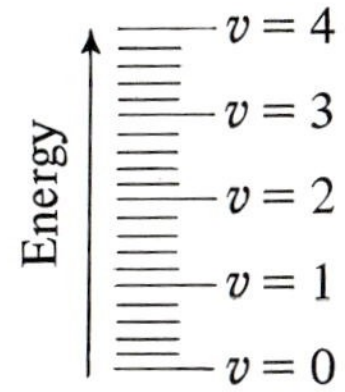

These levels are equally spaced purely for convenience. In fact as the vibrational quantum number v increases, the spacing diminishes until at some level they blend into a virtual continuum resulting in a 'smear' of energy states.

In the dark at room temperature, virtually all dye and pigment molecules are at the lowest vibrational level, $v = 0$, of the S_0 state and when a photon is absorbed, an electron is displaced into a higher orbital giving a singlet state $S_{1,2,3\ldots}$. The quantised vibrational and rotational energy levels associated with it are important because they result in a range of wavelengths being capable of causing excitation to each orbital. This governs the position and profile of the absorption band and the factors involved constitute what is known as the *Franck–Condon principle*, originally devised for diatomic molecules but equally applicable to complex organic molecules.

The Franck–Condon principle

When a molecule is in a specific vibrational energy level, the atoms vibrate about a mean internuclear distance where wave mechanics predicts the oscillator should spend most of its time. The transition of an electron into an anti-bonding orbital is much more rapid (about 10^{-15} s) than the period of vibration of the molecule (about 10^{-13} s) so that when the molecule enters the excited state the mean internuclear distance is unchanged. If this is the same as that of the excited molecule in its $v = 0$ vibrational level, then the most frequent transition will be from S_0 ($v = 0$) to S_1 ($v = 0$), which is that requiring the lowest amount of energy. This will result in the absorption band having a near vertical profile for its long wavelength boundary, a factor which leads to high chroma colours. In the case of most dyes and pigments, however, the mean internuclear distances are different for ground and excited states: the most frequent transition is therefore to $v = 1, 2, 3\ldots$ resulting in a less steep profile. The frequency to each vibrational energy level – which governs the whole profile of the absorption band – is determined by the vibrational wavefunctions for the two states, actually the magnitude of the overlap integral.

These effects are clearly illustrated in the absorption spectrum of lycopene (XXIII) shown in figure 2.3, the most intense band being that corresponding to excitation to the $v = 1$ level of the S_1 excited state. Lycopene and the related carotenoid pigments, however, are unusual among colorants in showing such fine structure as normally only a single peak appears for each excited state.

The absorption of photons and all of the mechanisms by which the molecule in an excited state loses its excess energy are best illustrated by a detailed energy level diagram such as that of figure 2.4. The initial absorption to three upper singlet states is shown on the left-hand side, only one photon absorption for each state being shown for the sake of clarity. Irrespective of the vibrational level reached by the initial absorption, the molecule loses excess vibrational energy within about 10^{-13} s to occupy the $v = 0$ level. The detailed mechanism of this loss is not clearly understood but apparently it proceeds through a series of collisions with other molecules rather than in one total interchange of energy. The excess energy therefore appears as heat, this radiationless transition being

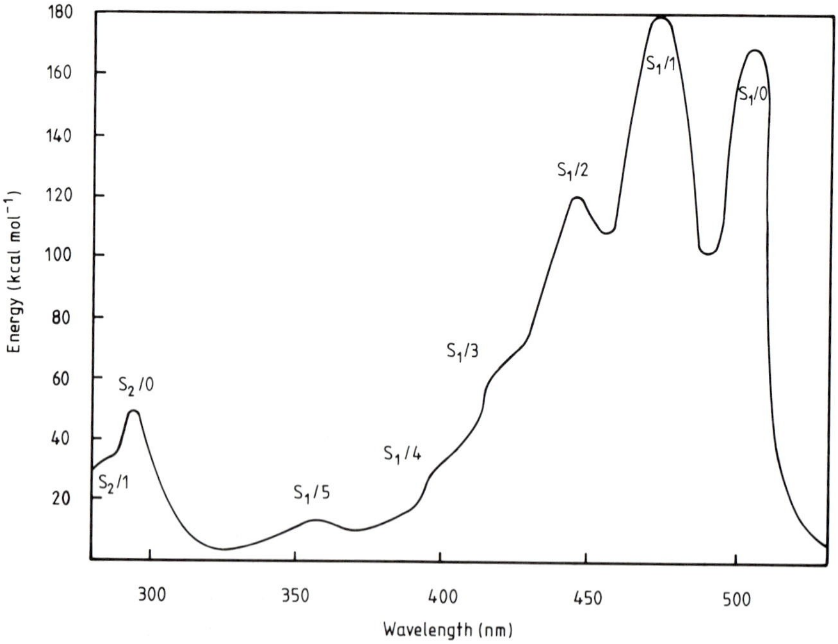

Figure 2.3 Absorption spectrum of lycopene in hexane.

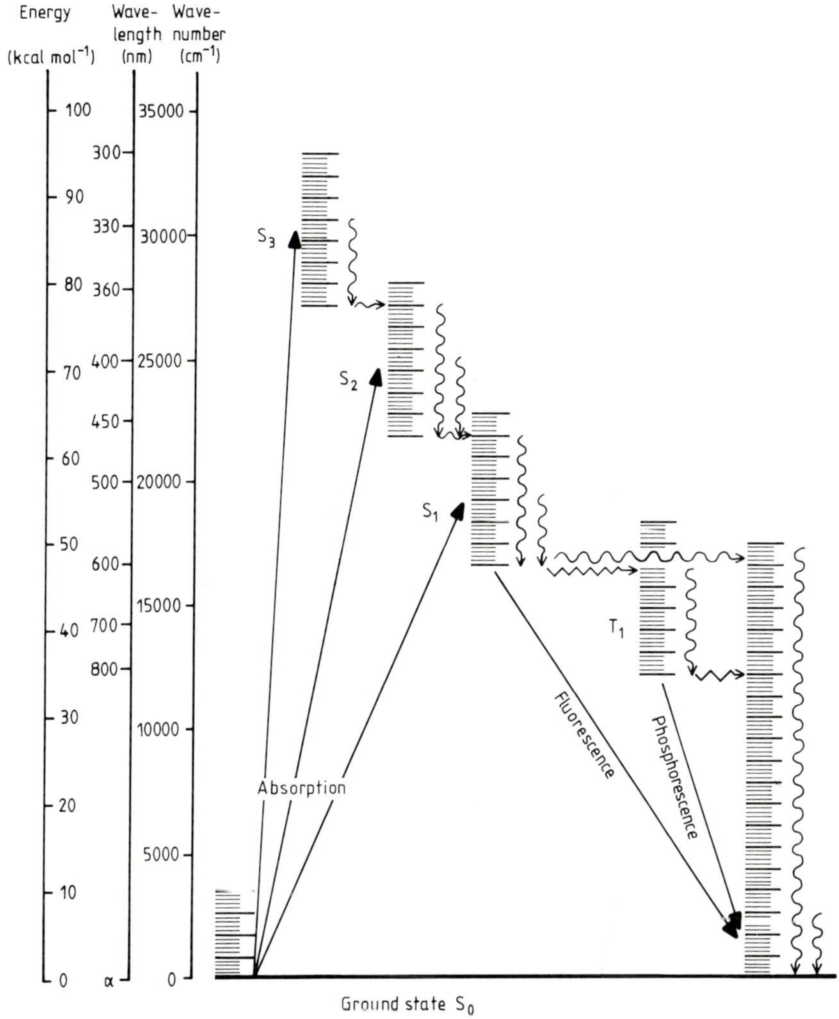

Figure 2.4 Energy level diagram (see text).

conveniently termed 'vibrational relaxation': it is represented by vertical wavy lines pointing downwards in figure 2.4.

When a molecule reaches the $v = 0$ level of the S_2 or higher singlet state it undergoes *internal conversion* to a high vibrational level of the next lowest singlet level. This is illustrated by a horizontal wavy line indicating no change in energy. The molecule then loses energy again to reach the $v = 0$ level, the overall effect being that, irrespective of the level of the singlet state reached by photon absorption, within about 10^{-11} s the molecule is at the lowest excited singlet state S_1, $v = 0$. This is the molecular equivalent of a geographical 'great divide' though there are three paths which the molecule can follow. Each of these will be described in detail. It is most convenient to start with the two best understood, fluorescence and intersystem crossing, ending with that least understood, internal conversion to the ground state.

Fluorescence

After a molecule has been at the S_1, $v = 0$ level for about 10^{-8} s there is a probability that it will lose its excess energy by a single burst of radiation as the electron in the π^* anti-bonding orbital returns to its original orbital. Again the Franck-Condon principle is invoked, this time it is the vibrational levels of the ground state which govern the profile of the emission band.

The emission band is centred at a longer wavelength than the absorption band, a fact discovered by Stokes as long ago as 1852. The reason for this is that part of the energy difference between the S_0, $v = 0$ level and the upper vibrational level of the S_1 state is accounted for by the vibrational relaxation occurring before and after fluorescent emission. In simple molecules such as anthracene, the spacing between the vibrational energy levels of the S_0 and S_1 states is virtually identical and the emission band is a faithful mirror image of the absorption band even though the latter has several peaks. In the case of fluorescent dyes and brightening agents, however, the emission band is usually very much narrower than the absorption band.

The extent to which the emission band is displaced to longer wavelengths depends on the difference in mean internuclear distances of the $v = 0$ levels of the two states and this is never large enough to prevent the absorption and emission bands overlapping.

Irrespective of the wavelength of excitation, the profile of the emission band is always the same. This explains why an ultraviolet lamp is so effective in revealing fluorescence even if the main excitation band is in the visible region, as it is in the case of dyes which are fluorescent.

As the emission and absorption bands overlap it is possible for light to be emitted at a shorter wavelength than that causing excitation, a phenomenon known as 'anti-Stokes emission'. This can be readily seen from the following simplified energy diagram. The most probable transitions are those shown by

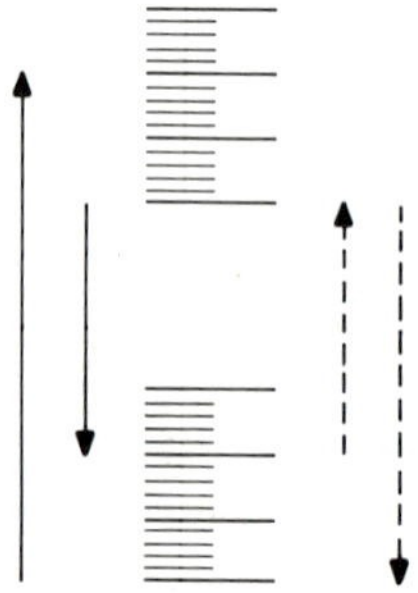

solid lines and that for emission is the shorter indicating less energy therefore longer wavelength. The transitions represented by broken lines are not, however, forbidden and will therefore occur, though infrequently. In this case the emission line is longer indicating a shorter wavelength.

The technological importance of fluorescence. When the emission band of a fluorescent dye lies wholly or partly in the visible region the effect on the perceived colour is potentially beneficial giving colours which are of a higher brilliance than non-fluorescent colours of the same hue. Many of the early synthetic dyes were fluorescent but these were very fugitive and when faster dyes began to be discovered at the start of the twentieth century, virtually none were fluorescent when dyed on the fibres available at that time. Fluorescence first began to be exploited with colourless dyes which absorbed in the near ultraviolet region (350–400 nm) and emitted blue light: the effect of this was *brightening.*

Fluorescent brightening agents. All natural textile fibres reflect somewhat less light at the violet–blue end of the spectrum than in the green–red region and this deficiency increases with age when the textile becomes distinctly yellowish. This yellowness was originally overcome by 'blueing'. This process involves a fine dispersion of the reddish-blue pigment, ultramarine, being added to the final rinse liquor both in laundries and in the domestic wash. This reduced the reflectance in the green–red region of the spectrum which resulted in the unattractive yellowness being replaced by a more acceptable greyness. In 1929, Krais found that if bleached flax was soaked in an aqueous extract of the husks of horse chestnuts, it appeared whiter or brighter. The extract contained a colourless glycoside of 6:7 dihydroxycoumarone which was fluorescent and when illuminated by a source containing ultraviolet light, emitted blue light which counteracted the residual yellowness of the fibre without reducing the total amount of light reflected (Krais 1929). Unfortunately, however, after only a short exposure to daylight the textile became even yellower than it was before treatment so the effect could not be exploited. In 1937 ICI patented the use of synthetic colourless, fluorescent dyes – derivatives of 4.4′ diaminostilbene 2.2′ disulphonic acid – for making hidden markings on, for example, banknotes. Under an ultraviolet lamp the markings became visible but the value of such compounds for brightening white textiles was not realised until 1940 when several were patented by the IG and by Lever Brothers. By the end of 1982, 378 fluorescent brightening agents (FBA) had been marketed which are not only incorporated into detergents but into white textiles and paper during manufacture. These compounds are also termed optical bleaches and fluorescent whitening agents.

Curves illustrating the effect of treating slightly yellowed cotton with a blueing agent and with an FBA are shown in figure 2.5. The presence of the FBA increases the total amount of light coming from the textile and this increases the

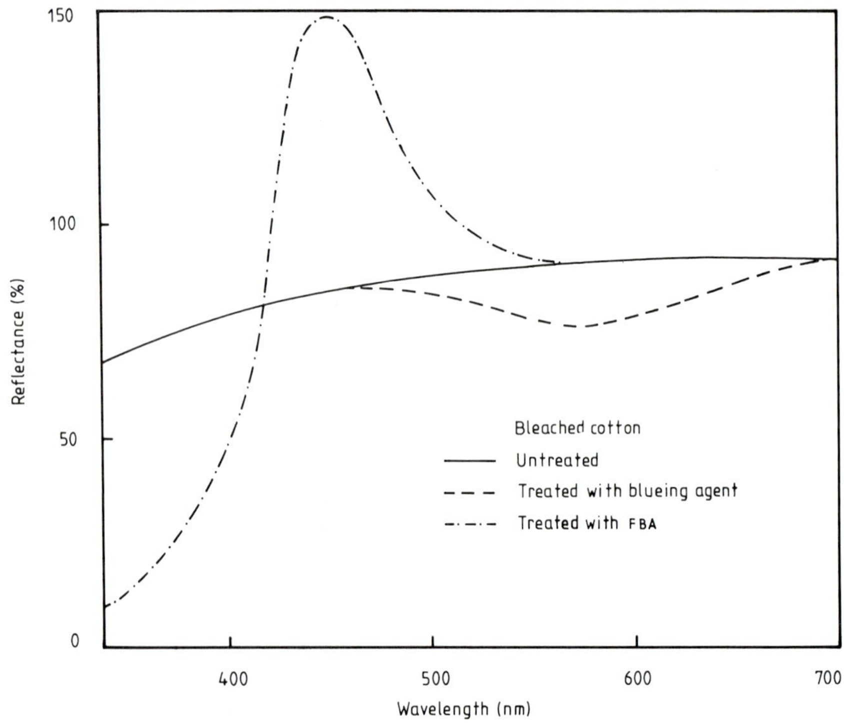

Figure 2.5 Spectral reflectance curves of various cotton specimens.

brightness as well as eliminating the yellowness. Beyond a certain concentration of FBA any further increase in brightness will be accompanied by the introduction of a distinct hue which is critically dependent on the wavelength of the peak of the fluorescent emission. When this is near the violet end of the spectrum, the white shade produced has a violet cast or tint which is normally described as being redder than a pure white, the more logical term 'violetter' never having become acceptable. When the emission peak moves towards the red end of the spectrum the tint of the white shade passes through a neutral blue at about 465 nm to greenish-blue whites, the preferred shade of white varying from country to country.

Fluorescent dyes. Though fluorescent brightening agents are unquestionably dyes, it is convenient to restrict the composite term 'fluorescent dye' to those which possess a strong absorption band in the visible region, i.e. which are coloured in the everyday meaning of the term.

As has already been mentioned, several of the early dyes were fluorescent, for example, fluorescein (CI Acid Yellow 73:45350) which has been, and still is, widely used to trace the course of underground streams and as a marker at sea. Their very low light-fastness on the fibres available at that time, however, limited their usefulness but during the 1940s, Joseph and Robert Switzer found that reasonable light-fastness was obtained on butylated urea-formaldehyde resins which could be ground to form printing inks and paints. One of the most useful dyes for this purpose was rhodamine B (CI Basic Violet 10:45170) discovered in 1887. This not only gave brilliant magenta colours on its own but by blending with the yellower rhodamine 5G (CI Pigment Red 81:45160:1) and with fluorescent yellow dyes a whole range of greenish-yellow, orange, scarlet, red and magenta shades were produced and became widely used under the trade name 'Day-Glo' for display purposes where their exceptional brilliance was an advantage (Ward 1972).

The combination of a greenish-yellow fluorescent dye and rhodamine B is interesting in that the wavelength of the greenish-yellow light emitted by the former coincides with the absorption maximum of the rhodamine and does not, therefore, emerge from the mixture. Though this light could have been absorbed by the rhodamine it is more likely that there is a direct, radiationless transfer of energy from the excited singlet state of the yellow to form the excited singlet state of the rhodamine which then fluoresces in the red.

Many new fluorescent dyes have been synthesised specially for synthetic fibres and are used to satisfy the demands of fashion and, much more important, to exploit the much greater visibility of such materials by incorporating them in safety clothing.

Fluorescence is traditionally detected – and exploited in places such as discos – by irradiation with ultraviolet light obtained cheaply by passing the light from a low-pressure mercury arc through Wood's glass. This consists of a broad band covering the range 320–400 nm. All fluorescent colorants possess at least one absorption band in this region and therefore irradiation is followed by fluorescent emission in the visible, preceded if necessary by internal conversion. This fact has undoubtedly contributed to the widely held belief, expressed in at least one recent textbook, that the brilliance of fluorescent dyes and pigments is caused by the ultraviolet component of daylight. Doubts about the truth of this assumption, were raised in the author's mind on seeing posters printed with 'Day-Glo' printing inks during a fog in Manchester in the days before it became a smoke abatement area. Their relative brilliance was unimpaired but the yellowness of the fog clearly indicated that the ultraviolet content of the daylight must have been virtually zero. The intensity of fluorescent emission is a product of the amount of exciting radiation at each wavelength and the degree of absorption at the same wavelength. In the case of fluorescent brightening agents there is no absorption in the visible region and therefore the whole of the fluorescent emission is caused by ultraviolet excitation. In the case of chromatic fluorescent colorants, however, the strongest absorption bands are usually those in the visible region and the intensity of radiation in the visible region is greater than that in the ultraviolet in the case of daylight so that the major part of fluorescent emission has been caused by visible light excitation. The negligible contribution of the ultraviolet component of daylight is illustrated by the curves showing the spectral radiance factors (see chapter 9) of a fluorescent orange printing ink irradiated with a source resembling daylight (figure 2.6(*a*)) and by the same source with the ultraviolet component removed (figure 2.6(*b*)). Such a difference corresponds to a perceived colour difference which is only about three times greater than just perceptible.

The negligible contribution made by the ultraviolet component of daylight to brilliance makes it feasible to apply a transparent coating containing an ultraviolet absorber to improve the light-fastness if the ultraviolet region contributes significantly to fading on exposure. Such specimens will not appear to be fluorescent if inspected under an ultraviolet lamp and the simplest method of detecting fluorescence in such cases is to measure the total spectral radiance factors using two different sources such as one simulating daylight and the other, a tungsten filament lamp. Total spectral radiance curves are shown in figures 2.6(*a*) and (*c*). If the specimen is not fluorescent the two curves will be virtually identical. Some spectrophotometers which employed a single source simulating daylight had a yellow filter which could be placed between the source and the specimen and this proved to be equally effective.

The fluorescence of green and blue dyes. Fluorescence is invariably associated with yellow, orange and red colorants and it was not until 1911 that it was discovered that many green and blue dyes were fluorescent (Pauli 1911); Ramsley (1967) lists 43 of them. The reason why their fluorescent properties escaped detection for so long is as follows. The emission band of all fluorescent colorants corresponds to the transition from the lowest excited singlet state S_1 to the ground singlet state S_0, even if the original excitation was to S_2, S_3.... Such emission peaks are invariably at a wavelength which is about 80 nm longer than the absorption peak corresponding to the $S_0 \rightarrow S_1$ transition. Green and blue dyes, however, must have such a band in the red region of the spectrum and if they are fluorescent the emission must lie mainly in the infrared. It will not, therefore, be detected by inspection under an ultraviolet lamp.

A substantial increase in brilliance of green shades can, however, be achieved by producing them with a combination of a bright blue colorant and a fluorescent yellow colorant. Some increases in brilliance of blue shades can be obtained by incorporating a fluorescent brightening agent with the colorant.

The fluorescence of azulene. Of the thousands of molecules whose fluorescent emission spectra have been determined, all except one correspond to emission from the lowest singlet state S_1 to the ground state S_0. The exception is the pale blue hydrocarbon azulene (XXIX) an isomer of naphthalene into which it is

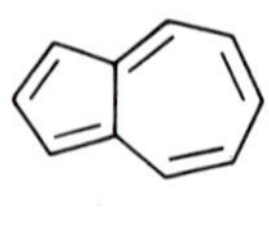

(XXIX)

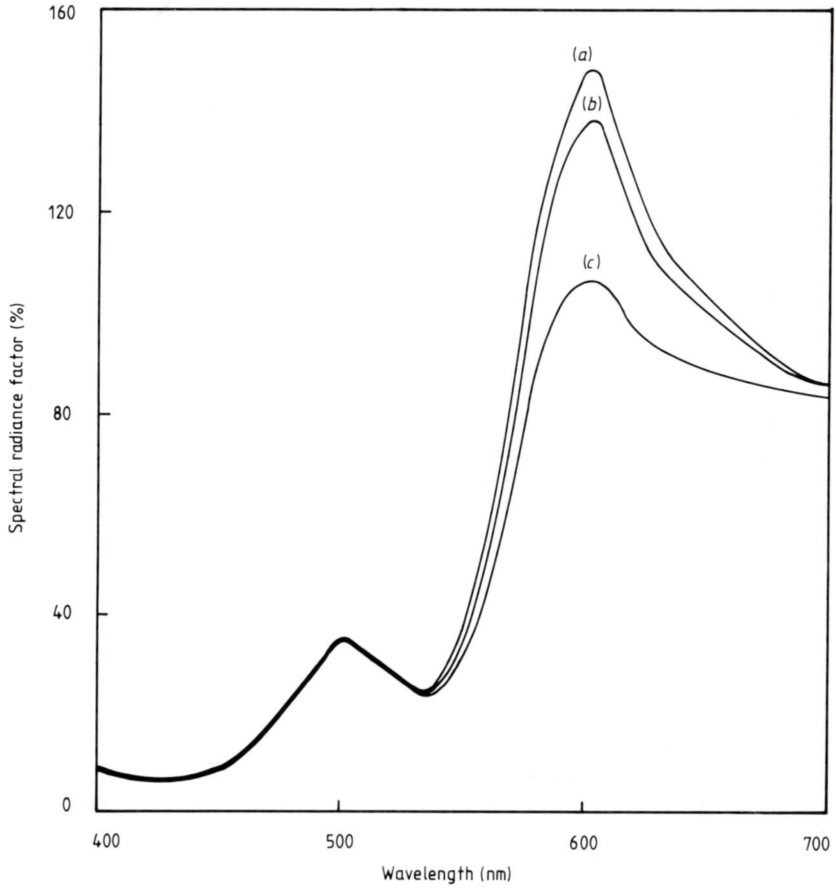

Figure 2.6 Spectral radiance curves of fluorescent orange printing ink. (*a*) Irradiance with source resembling daylight; (*b*) same source with ultraviolet component removed; (*c*) tungsten filament lamp.

transformed on heating to 270 °C. Azulene has an absorption band at 682 nm corresponding to the S_1 state and another at 357 nm (S_2) and when it is excited to the S_2 or higher levels, the fluorescence corresponding to $S_2 \rightarrow S_0$ alone is emitted and, moreover, if it is excited to the S_1 state, no emission can be detected (Beer and Longuet-Higgins 1955). The very large energy gap between S_2 and S_1 might be expected to prevent internal conversion but there are many other compounds with similar energy gaps which exhibit normal $S_1 \rightarrow S_0$ fluorescence. This does not, however, provide an explanation for the absence of $S_1 \rightarrow S_0$ fluorescence when excitation is to the S_1 state. Though of no apparent practical importance, azulene provides a fascinating glimpse of the simple questions which the relatively new science of photophysics has yet to answer. Another unanswered question is why there is no fluorescent dye in all the hundreds of thousands of azo dyes which have been synthesised.

After this extensive digression into the practical exploitation of fluorescence we must return to the main theme and consider the second pathway a molecule in the lowest excited singlet state may follow – intersystem crossing.

Intersystem crossing

When an electron is displaced into a π^* anti-bonding orbital it is unchanged in spin and hence the excited states resulting from photon absorption are also singlets. One of these electrons, now being the sole occupant of an orbital, may, however, reverse its spin. There are now two electrons unpaired in spin and they no longer make zero contribution to the spin angular momentum but a contribution which is a vector quantity having the components +1, 0 and −1. This produces three energy levels, though in complex molecules these are indistinguishable. The resultant electronic state is termed a *triplet.* It is surprising that the highly significant role of the triplet state in the spectroscopy of organic molecules only became apparent after the Second World War.

A transition from a singlet to a triplet state (or its converse) is described as intersystem crossing and this is illustrated by the zig-zag line from S_1, $v = 0$ in figure 2.4. The triplet state always has a lower energy than the corresponding singlet state (the molecular version of 'Hund's rule for atoms') and therefore intersystem crossing is to a higher vibrational level. Vibrational relaxation then occurs down to the $v = 0$ level in about 10^{-12} s at which point the molecule once again has three possible paths to the ground state – phosphorescence, slow fluorescence or reverse intersystem crossing.

Phosphorescence. Phosphorescence is similar to fluorescence in that the excess energy is lost in a single burst of radiation but it differs in two respects – the phosphorescence band being centred at a longer wavelength and its appearance being considerably delayed. The first of these differences is apparent from figure 2.4: the triplet state is always at a lower energy level than the corresponding singlet state. The second arises because radiative transitions between singlet and triplet states are 'forbidden' by the laws of quantum mechanics. This effectively prevents the absorption of a photon to give an $S_0 \rightarrow T_1$ transition but does not prevent a radiative transition in the reverse direction, i.e. phosphorescence, but it does delay it by periods of up to a second.

Slow fluorescence. The path of slow fluorescence is not shown in figure 2.4 for the sake of clarity but it can be readily traced. The molecule in the T_1, $v=0$ state absorbs thermal energy to reach the level of the S_1, $v = 0$ state and one of the unpaired electrons then reverses its spin causing an intersystem crossing in the reverse direction back to the singlet state which is followed by fluorescence. The profile of this fluorescent emission is identical with that of normal fluorescence but the indirect route delays its appearance and it is therefore termed 'slow fluorescence'.

Either phosphorescence or slow fluorescence was first observed for organic molecules in 1895. When fluorescent dyes were dissolved in water, the luminescence ceased as soon as the exciting radiation was switched off. When gelatin was added before irradiation an afterglow was observed. Parker and Hatchard later discovered that both phosphorescence and delayed fluorescence occurred after irradiating solutions of eosine (Parker and Hatchard 1961). This indicates that delayed luminescence can be expected to occur when fluorescent dyes are applied to substrates such as polymethylmethacrylate and polyacrylonitrile. Such combinations will cause problems when spectral radiance factors are being determined using spectrophotometers which irradiate the specimen with a short pulse of radiation from a xenon arc and measure the reflectance and emission within a few microseconds.

Reverse intersystem crossing. This is shown by the zig-zag line from the T_1, $v = 0$ level to an upper vibrational level of the ground state. Another line also leads to a higher vibrational level of the ground state – from S_1, $v = 0$, i.e. the third path from the 'great divide'. Although this involves internal conversion rather than intersystem crossing, the two are best considered together as radiationless transitions to the ground state.

Radiationless transitions to the ground state

Unfortunately the transitions $S_1 \rightsquigarrow S_0$ and $T_1 \rightsquigarrow S_0$ cannot be detected directly: all one can do is to determine the quantum efficiency of each of the other alternative pathways to the ground state and if these do not sum to 1, the difference is a measure of the probability that a radiationless transition has occurred.

The probability that $S_1 \rightsquigarrow S_0$ transitions occur at room temperature in organic solvents has been assessed by Calvert and Pitts in their invaluable book *Photochemistry* (Calvert and Pitts 1966). They combined the quantum efficiency of triplet state formation $S_1 \rightsquigarrow T_1$, from the unpublished data of Lamola and Hammond with the quantum efficiency of fluorescence $S_1 \rightarrow S_0$ as collated by Bowen for a number of hydrocarbons. They concluded that radiationless internal conversion from the first excited singlet state to the ground state 'may be a major process for many aromatic molecules' (Calvert and Pitts 1966, p. 311). Whether this will apply to colorants or not depends on the probability of triplet state formation. If this probability is high, as it is for aceto- and benzo-phenone, then the probability that radiationless internal conversions will occur will be virtually zero.

The probability that $T_1 \rightsquigarrow S_0$ transitions occur at room temperature can only be a matter of speculation. Facts are available only for colourless compounds at the temperature of liquid nitrogen and the results obtained by different investigators are contradictory. Ermolaev and Robinson and Frosch concluded that it is a major process whilst Kasha expressed the view that in general radiationless transitions in aromatic compounds in rigid solvents at low temperatures do not occur (Ermolaev 1963, Robinson and Frosch 1963, Kasha

1960). It is not surprising that Calvert and Pitts described the role of radiationless transitions as 'one of the most controversial and significant questions currently facing molecular spectroscopists and photochemists'.

The frequency of photophysical processes in daylight. All of the numerous pathways just described lead the molecule back, unchanged, to the ground state so that these physical processes are cyclical. It is interesting to consider how frequently a dye molecule undergoes such a process when applied to a textile fibre and then exposed to the most intense radiation it is likely to encounter under normal circumstances, i.e. noon summer sunlight. Hemmendinger has undertaken the necessary calculations (Hemmendinger 1972) and concluded that:

(*a*) a molecule, on average, absorbs a photon only once every 15 s;

(*b*) if the molecule is neither fluorescent nor phosphorescent, only about one molecule in every thousand million is in an excited state at any one instant.

2.3 PHOTOCHEMICAL PROCESSES FOLLOWING LIGHT ABSORPTION

The photophysical processes which follow light absorption result in the absorbed energy being converted harmlessly into heat but the excited states formed following absorption can lose their excess energy by chemical reactions resulting in a loss of or change in colour. This extremely important phenomenon is known as 'fading'. As tanning of the human skin is only caused by exposure to ultraviolet light it was natural to assume that it was this component of daylight which caused fading and transparent screens containing UV absorbers were widely used over shop windows containing knitting wools, for example. These were, however, of negligible value and were sometimes replaced by yellow screens which absorbed violet and blue light as well. These were somewhat better but did not prevent fading during normal periods of exposure in a shop window because dyes of only moderate resistance to light are primarily faded by the visible light they absorb (McLaren 1955).

Extensive investigations into the chemical reactions involved in dye fading have been carried out but as the mechanisms are both highly specific and extremely complex no attempt will be made to summarise them here. Interested readers are therefore referred to the excellent comprehensive reviews which have been published (Egerton and Morgan 1970, Meier 1971, Griffiths 1972, Bentley *et al* 1974 and Leaver 1980).

A lesser-known photochemical reaction involving dyed fibres exhibits much less specificity though much complexity and this will be reviewed as it forms a very interesting chapter in photochemistry. This reaction is the photochemical degradation of the textile fibre catalysed by the presence of certain dyes of high light fastness, a phenomenon known as 'tendering'.

Tendering

It will be recalled that one of the landmarks of the synthetic dye era was the synthesis of alizarin (I) and indigo (II). In 1901 Rene Bohn, a Swiss employed by the German dye firm BASF, provided a perfect example of the 'enlightened empiricism' approach of dye chemists. Bohn attempted to make the anthraquinonoid analogue of indigo from the glycine derived from 2-aminoanthraquinone. He obtained a blue compound but when he attempted to vat it the expected colour change to pale yellow – characteristic of all the indigoid vat dyes – did not occur. Legend has it that he gave vent to his feelings of disappointment by putting the beaker down so forcibly that it cracked and the blue liquid ran over his white lab-coat. He instructed his assistant to wash it but and hour later the 'stain' was still there! The product he had produced was a vat dye, but of an entirely new chemical class, one characteristic of which being that their leuco compounds were not usually pale yellow in colour. The structure of this dye is as in (XXX). It was named *indanthrone* – CI Vat Blue 4:69800. When highly coloured automotive finishes were required in the 1950s indanthrone was found to have valuable properties and it was marketed as a pigment (CI Pigment Blue 60:69800).

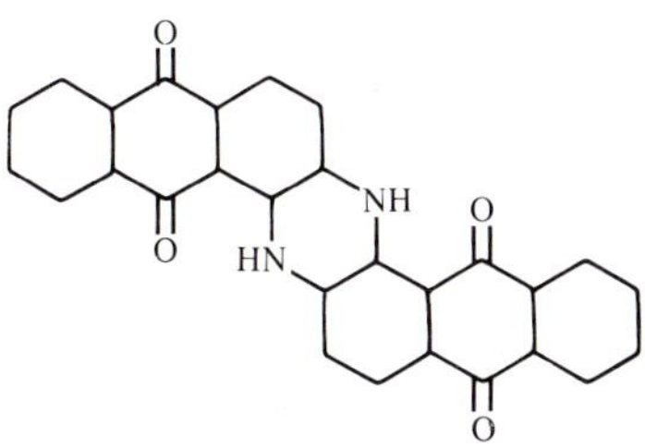

(XXX)

Dyes of this class had such excellent all-round fastness properties that fabrics dyed or printed with them were sold with a guarantee that they would be replaced if they faded in use (Morton 1929). It was soon discovered that curtains printed with certain of these dyes, which showed no sign of fading when first taken down, went into holes on washing, the outline of the holes corresponding exactly to areas of the fabric which had been printed with specific yellow and orange dyes. Haller showed, in 1924, that these dyes promoted the formation of oxycellulose on exposure to light, this degradation product being soluble in hot, alkaline soap solutions (Haller 1924). This phenomenon is usually known as *tendering* (fibre degradation caused by photo-oxidation). In 1929, Schofield and Goodyear showed that when cotton dyed with a tendering dye was exposed to light under moist conditions hydrogen peroxide was formed (Schofield and Goodyear 1929).

One of the most surprising features about tendering and non-tendering dyes was that activity seemed to be entirely unrelated to chemical structure. Venkataraman (1952) gives an excellent example of this: (XXXI) (CI Vat Yellow 26:

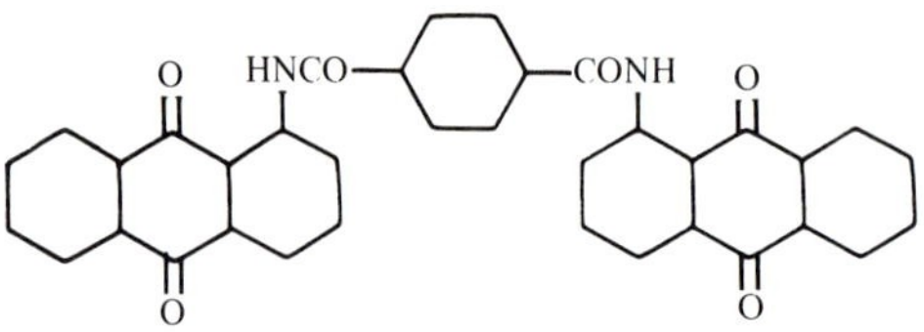

(XXXI)

65410) and (XXXII) (CI Vat Yellow 3:61725) are both active tenderers and it would be reasonable to associate this with the presence of the benzamido groups ($C_6H_5CO{\cdot}NH$—) if it were not for the fact that (XXXIII) (CI Vat Yellow 13:65425) is a non-tenderer.

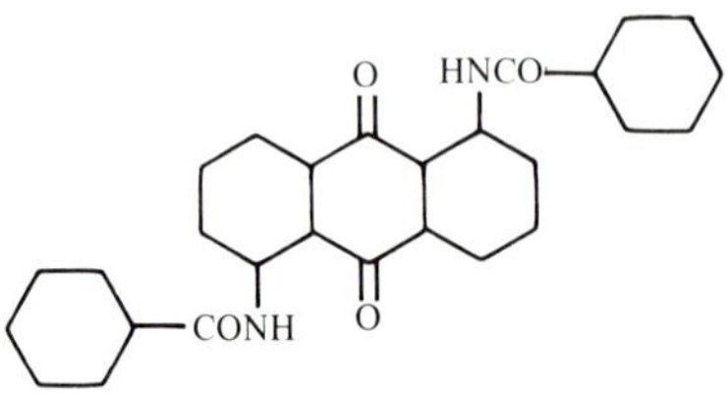

(XXXII)

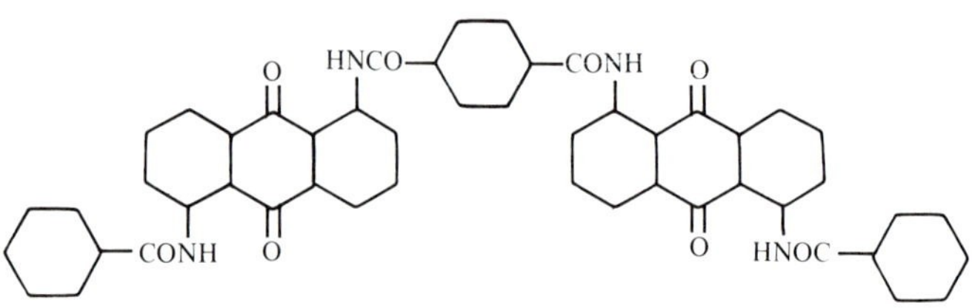

(XXXIII)

In the late 1940s, Egerton made a thorough investigation into this phenomenon and found that it was by no means confined to specific yellow and orange vat dyes on cellulosic fibres. Various direct, acid, basic, sulphur, vat and disperse dyes of many different colours sensitise the photochemical degradation of cotton, viscose rayon, silk, nylon and acetate fibres but only in the presence of oxygen, the effect usually being increased in the presence of water vapour. Zinc oxide and anatase (TiO_2) behave like sensitising dyes. Egerton confirmed that hydrogen peroxide was formed when oxygen and moisture were present and found that this was capable of diffusing several millimetres to degrade an undyed fibre or to oxidise starch-iodide paper. He also found that the oxidation of cellulose by hydrogen peroxide was increased by exposure to light (Egerton 1948).

In order to explain the mechanism of hydrogen peroxide formation Egerton invoked a theory developed by Kautsky (1938) to explain a photo-oxidation effect he had discovered. He absorbed the fluorescent yellow dye, acriflavine, on to silica gel and mixed it with silica gel coated with colourless leuco-malachite green. On irradiation *in vacuo* the acriflavine exhibited both fluorescence and a slow fluorescence lasting many seconds at room temperature and the leuco-malachite green remained colourless. When he admitted a trace of oxygen into the vessel, the fluorescence was quenched and the leuco-malachite green turned greenish blue. Kautsky's explanation was that the energy of an excited acriflavine molecule was transferred to an oxygen molecule promoting it into an excited state which then diffused over to adjacent leuco-malachite green and oxidised it to the bluish green dye, the acriflavine thereby returning to the ground state and becoming available for further, catalytic oxidation.

Photochemists had ignored his theory because experimental evidence that excited oxygen had been formed was lacking even though no satisfactory alternative mechanism had ever been put forward to explain the facts even by as late as 1960 when similar phenomena were observed (Rosenberg and Shombert 1960). Egerton, however, considered that this was the most likely first stage in phototendering and in moist air the second stage would be the formation of hydrogen peroxide by a reaction between oxygen in an excited state and water vapour (Egerton 1949). He had, however, overlooked the fact that Kautsky had stated that the influence of water on the photo-oxidation was an adverse one and Bamford also pointed out that there was no evidence that excited oxygen can form hydrogen peroxide by reaction with water (Bamford 1949).

Bamford's comment was made at a symposium entitled *Photochemistry in Relation to Textiles* organised by the Society of Dyers and Colourists and in another paper Bamford and Dewar put forward an alternative mechanism to Egerton's 'activated oxygen' theory. They postulated that the dye in an excited state reacted with the substrate by a process of hydrogen atom abstraction which created a 'free-radical centre' which is then attacked by molecular oxygen (Bamford and Dewar 1949). We shall see that the conflict between the 'hydrogen abstraction' theory and the 'activated oxygen' theory is probably not resolvable. At this symposium Bowen and Livingstone independently drew attention to the undoubted importance of the triplet state (Bowen 1949, Livingstone 1949) whose nature had been finally established only five years earlier (Lewis and Kasha 1944).

The triplet state was expected to be involved in many photochemical reactions because, like the excited singlet state, it is highly reactive, a molecule in an excited state possessing two orbitals each containing only one electron, a state the organic chemist describes as a *di-radical*. The triplet state would be expected to be much more effective, however, because its lifetime is known to be several orders of magnitude longer. The quantum efficiency of dye fading is extremely low; if a dye had a quantum efficiency of 1 it would be completely faded after less than ten seconds exposure to sunlight (Giles *et al* 1972). The most fugitive commercial dye would require several hours exposure. Bowen suggested that this low efficiency might be because of the low probability of the triplet state being formed (Bowen 1949).

Shortly after this symposium, the British Rayon Research Association (BRRA) began an intensive study of the fundamental mechanisms involved in tendering. Tendering is a more serious problem with viscose rayon than with cotton because the undyed fibre has a lower tensile strength, especially when wet. The BRRA believed that if these mechanisms could be elucidated, methods of minimising tendering might be devised. They began their studies with 'model systems' in which a simple anthraquinone instead of a complex vat dye was used as the sensitiser and a simple alcohol instead of complex cellulose as the acceptor. This technique was first used by Bolland and Cooper and gave significant insight into the mechanism of tendering. The sensitiser in an excited state reacted with ethanol to form a semiquinone radical, subsequent reactions with oxygen and ethanol leading to the oxidation of the latter (Bolland and Cooper 1954). Wells showed that this primary process was one of hydrogen abstraction and not electron transfer (Wells 1956). Cooper later found that substituted anthraquinones can be divided broadly into sensitisers and non-sensitisers according to the position of methyl, chloro- and sulphonic acid groups (unpublished, but quoted by Bridge 1960a).

The most valuable technique introduced by the BRRA was the Nobel Prize-winning technique of flash photolysis (Norrish and Porter 1949). An intense flash of visible and near UV light caused the initial formation of excited singlet states followed by intersystem crossing to form the lowest excited triplet state whose lifetime was long enough to permit an absorption spectrum to be obtained by means of a second, much weaker, flash; subsequent 'analytical' flashes could then be used to measure the rate of decay of the triplet state. Any bands appearing after flashing would be caused by excitation from the T_1 state to T_2, T_3 etc (which are not shown in figure 2.4) or possibly from a subsequently formed radical or radical ion. The first of these studies used liquid paraffin instead of ethanol as its high viscosity prolonged the triplet lifetime: the sensitiser was tetramethyl-p-benzoquinone. To everyone's surprise the kinetics of the system indicated that the semiquinone radical was being formed from the singlet state and not the triplet (Bridge and Porter 1958).

Although Egerton had found that photochemical degradation of textile fibres could be caused by red, green and blue dyes the fact that the majority – and all of the technically important tendering dyes – were yellow or orange in colour suggested that a study of their absorption spectra should lead to an explanation of the mechanisms involved. It was complicated, however, by the fact that there were many non-tendering yellow and orange dyes whose principle absorption bands had similar positions and shapes to those of tendering dyes. In 1933, Preston observed that tendering dyes had, in addition, another band in the ultraviolet, i.e. in the 360–400 nm region (unpublished, but quoted by Schofield and Turner 1933). In 1954 Cooper observed that sensitising anthraquinone derivatives had a shoulder on the long wavelength side of the principle absorption band (Cooper private communication). Moran and Stonehill examined the spectra of 39 anthraquinonoid vat dyes and related compounds and also found that several of the active compounds exhibited weak bands in the 400 nm region but that none of the inactive dyes did so though they admit these may have escaped detection (Moran and Stonehill 1957).

One of the first attempts to classify the electronic spectra of complex molecules was made by Burawoy who classified the weak, long-wavelength absorption bands of compounds containing —CO, —NO, —N=N— and other hetero groups as R-bands, which he thought were due to direct transitions from the ground state to an excited triplet state (Burawoy 1939). Moran and Stonehill therefore attributed the weak bands they had observed in tendering dyes to direct triplet state formation but by 1958 it had become generally accepted that the vast majority of these R-bands did not involve the triplet state but were transitions to an n-π^* singlet state. In 1960, Bridge expressed the view that such a transition would be expected to result in tendering rather than a π-π^* transition for the following reason: when cellulose is dyed with vat dyes there is a strong possibility of hydrogen bond formation (Venkataraman 1952, Peters and Sumner 1955) and the group in the dye which would be involved is that possessing an n electron which if it were displaced into a higher orbital might be expected to produce a reaction (Bridge 1960b).

The subsequent history of the elucidation of the mechanism of tendering can best be followed if the two major aspects are considered separately even though many of the investigations covered both, i.e. the specific chemical reactions occurring and the nature of the excited state involved.

The chemical reactions causing tendering

Photochemists, with the sole exception of Egerton, had shown complete hostility to the activated oxygen theory of Kautsky from the date of its inception, 1939, until the period 1963-4 when the situation changed dramatically. The cause of the change was the elucidation of the mechanism underlying the observation, first made as long ago as 1927, that the decomposition of hydrogen peroxide was often accompanied by the emission of red light (Khan and Kasha 1963, 1964a, 1964b, Foote and Wexler 1964, Corey and Taylor 1964). The decomposition resulted in the formation of oxygen molecules in an excited state and the red light was emitted as these molecules underwent a complex radiative transition to the ground state (Arnold *et al* 1964). Molecular oxygen is 'somewhat unique' among molecules containing an even number of electrons in that the ground state configuration is that of a triplet. The two excited states commonly occurring are both singlets, the one of higher energy (38 kcal) being designated $^1\Sigma$, that of lower energy (22 kcal) being designated $^1\Delta$.

The connection between this discovery and phototendering is that when singlet oxygen generated chemically from hydrogen peroxide was used to oxidise a number of organic compounds each of which gave a highly specific group of end-products, exactly the same products were obtained using the process of dye-sensitised photo-oxidation by gaseous oxygen. Kautsky's theory was thus drawn back into the mainstream of photochemical research and has developed at a rapid pace ever since. As Kearns has remarked 'the range of phenomena, reactions and systems in which singlet oxygen is now believed to be directly or indirectly involved is indeed remarkable' (Kearns 1971).

The first proof that some tendering vat dyes generated singlet oxygen when exposed to light in the presence of molecular oxygen was obtained by Griffiths and Hawkins (1973). Four yellow/orange vat dyes varying in activity from borderline to severe were irradiated in solution and on cotton containing tetraphenylcyclopentadienone (TPC) which reacts extremely readily with $^1\Delta$ oxygen to give cis-dibenzoylstilbene (CDS). The most comprehensive data on tendering was obtained by Egerton (1947) who exposed specimens of cotton dyed with 36 vat dyes to sunlight and determined the increase in fluidity, one of the standard methods of determining the extent of tendering. Their results are given in table 2.2. These studies were later extended by Garston using dyed fibres only: his results are given in table 2.3 (Garston 1980).

Table 2.2 Generation of cis-dibenzoylstilbene by vat dyes exposed to light (Griffiths and Hawkins 1973).

Dye	Rank order (Egerton)	Conversion to CDS (%)	
		Solution	Fibre
CI Vat Yellow 2:67300	2	18	36
CI Vat Yellow 26:65410	6	94	–
CI Vat Orange 9:59700	7	100	14
CI Vat Yellow 1:70600	11	2	4
Undyed cotton	18	–	4

Table 2.3 Generation of cis-dibenzoylstilbene by vat dyes exposed to light (Garston 1980).

Dye	Rank order (Egerton)	Conversion to CDS (%)
CI Vat Yellow 21:69705	1	58
CI Vat Yellow 2:67300	2	50
CI Vat Orange 21:69700	5	33
CI Vat Orange 9:59700	7	17
CI Vat Yellow 1:70600	11	10
Undyed cotton	18	8
CI Vat Green 1:59825	24	8
CI Vat Blue 20:59800	–	6
CI Vat Blue 12:69840	–	4

Bentley *et al* had, however, previously criticised Griffiths and Hawkins' conclusion on the grounds that whilst tendering dyes undoubtedly do produce singlet oxygen, it does not necessarily follow that it is this which actually causes tendering. Singlet oxygen production indicates the formation of a long-lived excited state but this could react either with oxygen or with the fibre by hydrogen (or electron) abstraction with tendering resulting from the latter rather than the former reaction (Bentley *et al* 1974). These investigators were members of a team at Salford led by Phillips who were the first to apply flash photolysis to cellulose films dyed with tendering vat dyes. This technique overcomes the criticisms often raised by Egerton against the evidence from model compounds, namely that they do not simulate the conditions occurring in practice (Egerton and Morgan 1971). Davies *et al* flashed in air cellulosic films dyed with CI Vat Yellow 2:67300 and CI Vat Yellow 4:59100 (ranking order of tendering being 2 and 3 respectively). The spectra they obtained resembled those from the flash photolysis of the same dyes in alcoholic solutions which indicated that hydrogen abstraction had occurred with Vat Yellow 2 and electron transfer with Vat Yellow 4 (Davies *et al* 1972, 1973a).

Convincing evidence as to the involvement of singlet oxygen as a cause of and not as a by-product of tendering should be obtained by determining if a quencher of singlet oxygen protects dyed cellulose against tendering. A very effective quencher is furan vapour not only because it reacts very readily with singlet oxygen, but also because a reaction in the gas phase would undoubtedly be favoured over a reaction at a gas–solid interface. Davies *et al* irradiated fabrics dyed with CI Vat Yellow 2 and CI Vat Yellow 4 in a stream of oxygen with and without the presence of furan vapour: no significant protection was given by the singlet oxygen quencher which would seem to be proof that singlet oxygen is *not* the cause of tendering with these two dyes (Davies *et al* 1973b). However, other singlet oxygen quenchers have been applied to cotton dyed with Vat Yellow 2 which did afford some protection. Garston obtained a 12% reduction in degradation using TPC (Garston 1980) and Griffiths and Hawkins found an increase in fluidity of only 5–11% when films were impregnated with diazabicyclo (2, 2, 2) octane compared with a 22–23% increase in its absence (Griffiths and Hawkins 1976).

Opponents of the singlet oxygen theory argued that there was no evidence that singlet oxygen could oxidise cellulose but in their 1976 paper Griffiths and Hawkins described experiments which showed that it could, and at a rate comparable with that of an efficient hydrogen abstractor. It is somewhat ironic that the authors thought that the reaction between singlet oxygen and cellulose was most likely to be one of hydrogen abstraction! Though this paper would seem to have removed the last remaining obstacle to the acceptance of singlet oxygen being a cause of tendering the authors concluded it with the following pertinent comment: 'it is important in interpreting experimental results that one mechanism should not be considered to the exclusion of the other, as has often been done in the past'. This view was later endorsed by Garston (Garston 1980) and by Allen and McKellar who added 'it is doubtful whether the relative importance of the "singlet-oxygen" and "hydrogen atom or electron abstraction" theories of phototendering will be resolved' (Allen and McKellar 1980).

In both mechanisms the dye behaves as a catalyst for phototendering by the oxygen in the air. When the singlet oxygen mechanism is operating, the dye returns to the ground state when it transfers its excess energy to an oxygen molecule

$$^3D^* + O_2 \rightarrow D + {}^1(O_2)^*$$

thus becoming available for further singlet oxygen formation until it has been destroyed by a secondary photochemical reaction – fading. When tendering is caused by direct reaction between the excited dye and cellulose the dye is converted into a semiquinone radical if the reaction is one of hydrogen abstraction or into a semiquinone radical anion if the reaction is one of electron transfer. In both cases the dye is re-formed by oxidation by atmospheric oxygen though the possibility of singlet oxygen being involved cannot be ruled out.

Different mechanisms have been proposed to explain the formation of hydrogen peroxide when cellulose is phototendered in moist air. If the mechanism is one of hydrogen (electron) abstraction, subsequent oxidation of the semiquinone radical or radical-ion forms $HO_2\cdot$ radicals which then combine

$$HO_2\cdot + HO_2\cdot \rightarrow H_2O_2 + O_2.$$

If the mechanism involves singlet oxygen Egerton and Morgan postulated the formation of an ion-pair complex between the dye in its triplet state and oxygen

$$^3D^* + O_2 \rightleftharpoons \cdot D^+ \text{-------} O_2^-\cdot.$$

This then reacts with hydroxyl ions followed by hydrogen ions to give $HO_2\cdot$ radicals

$$\cdot D^+ \ldots\ldots\ldots O_2^- \rightarrow \cdot D^+ \ldots\ldots\ldots OH^- + O_2^-\cdot$$

$$O_2^-\cdot + H^+ \rightarrow HO_2\cdot$$

$$HO_2\cdot + HO_2\cdot \rightarrow H_2O_2 + O_2$$

(Egerton and Morgan 1971) but this hypothesis has never been substantiated.

The excited state involved in tendering

Bridge and Porter's conclusion that it was the singlet state of tetramethyl-p-benzoquinone which reacted in liquid paraffin was shown in 1968 to have been incorrect, one of the transients observed having been wrongly assigned (Wilkinson *et al* 1968). There is now unanimous agreement among photochemists that by far the great majority of photochemical reactions are initiated by the triplet state of lowest energy.

The nature of this state, whether it is n-π^* or π-π^*, would be expected to be important because Porter and Suppan found that the photochemical reactivity of substituted benzophenones was about ten times greater when the lowest triplet was n-π^* than when it was π-π^*. This displacement of the n electron into a π^* orbital resulted in an electron deficiency on the carbonyl group which promoted hydrogen abstraction or electron transfer reactions. If a π-π^* transition was accompanied by an intramolecular charge-transfer (C-T), the increase in electron density on the carbonyl group reduced the reactivity. They assessed the relative reactivities of the n-π^*, π-π^* and C-T states as being 100:10:1 (Porter and Suppan 1965). This would account for Cooper's observation that sensitising anthraquinone derivatives had a shoulder on the long wavelength side of the principle absorption band as if this were due to a weak n-π^* band (which is likely) the lowest triplet would be expected to be n-π^*. Schorer and Dörr also concluded that tendering dyes did have n-π^* bands at longer wavelengths. However, their evidence for this was not particularly convincing as the most active in their group of active dyes was only rated seventh in Egerton's list and their group also included CI Vat Yellow 29:68400 and CI Vat Yellow 31:68405, the first greenish-yellow *non-tendering* vat dyes to be marketed (Schorer and Dörr 1964). Moran and Stonehill's observation that several active vat dyes had weak bands in the 400 nm region, however, does not suggest that the lowest triplet state will be n-π^* in character. The reduction in energy on passing from a singlet to a triplet state, the so-called singlet-triplet splitting, is always much greater for π-π^* than for n-π^* states so that as an n-π^* band at 400 nm will have a higher energy level than the principle π-π^* band (at 450 nm or greater), the n-π^* triplet must be at a higher energy level than the π-π^* triplet.

Davies *et al* investigated the spectroscopic properties and photochemical behaviour of two of the most active vat dyes, Vat Yellows 2 and 4. Vat Yellow 2 readily reacted by both hydrogen abstraction and electron transfer according to pH but Vat Yellow 4 reacted only by electron transfer (Davies *et al* 1972). Though they originally believed that the first excited singlet state (and therefore the lowest triplet) of both dyes was π-π^* in character, it is extremely difficult to identify the character of an excited state of a vat dye, partly because they are usually insoluble in diagnostic solvents and partly because any weak n-π^* band is likely to be swamped by a much stronger π-π^* band. Davies, therefore, now believes that in the case of Vat Yellow 2 the lowest triplet may have been n-π^* which would completely account for its photochemical behaviour. The first excited state of Vat Yellow 4, however, was clearly π-π^* showing considerable intramolecular charge-transfer character which would explain its negligible reactivity towards hydrogen abstraction. Electron transfer to the carbonyl group would also be prevented but Davies believes that efficient electron transfer occurs to the ring system which intramolecular charge-transfer has made electron-deficient (Davies, private communication).

Two of the three types of triplet state have therefore been shown to cause photodegradation and the third, π-π^*, would be expected to be involved if tendering is ever caused by photosensitised activated oxygen. This is not because such triplets give a higher quantum yield of singlet oxygen than n-π^* triplets – there is no theoretical reason why there should be any difference – but because the lifetimes of π-π^* triplets are about one hundred times longer than those of n-π^* triplets both in solution and in frozen glasses. The lowest triplet state is most likely to be π-π^* for dyes other than yellows because they have at least one π-π^* absorption band at longer wavelengths. It is surely no coincidence that among the most efficient sensitisers of singlet oxygen are rose bengal (CI Acid Red 94:45440), methylene blue (CI Basic Blue 9:52015) and rhodamine B (CI Basic Violet 10:45170). These dyes are not tenderers solely because they will have faded long before fibre degradation has occurred.

The nature of the triplet state, therefore, does not appear to be critical: a dye of high light fastness is likely to cause tendering if it passes into a long-lived triplet state. Conversely, it will not cause tendering if the lowest excited singlet state is deactivated before intersystem crossing to a triplet state can occur as Bowen (1949) had predicted or if the triplet so formed is rapidly deactivated. There is considerable supporting evidence for this view. In hydrogen abstraction reactions Millich and Oster found that the only acridine dyes which were not reduced by allylthiourea on irradiation were those which were not phosphorescent, i.e. which did not pass into the triplet state (Millich and Oster 1959). Dearman and Chan (1965) found that among twenty substituted anthraquinones, those which did not undergo photoreduction were those which did not phosphoresce. Garston found that a decrease in tendering activity was accompanied by a reduction in the lifetime of the lowest triplet state as the existence of these states could not be detected by electron spin resonance in the case of the least active yellow/orange dyes they studied (although they did phosphoresce) unlike the three inactive blue/green dyes (Garston 1980). Zweig and Henderson found that many dyes of different colours and chemical constitutions were efficient sensitisers for singlet oxygen with the exception of the four azo dyes they investigated – they attributed their inactivity to the ‘exceptionally short triplet lifetime’ (Zweig and Henderson 1975).

3

The Application of Colouring Matters

Colouring matters are conveniently divided into dyes and pigments although, as we have already seen, some colorants have been used for both purposes, indigo being the classical example. Many synthetic colorants are also marketed as both pigments and as dyes.

Pigments

The basic definition of a pigment is that it is a colorant in particulate form which is insoluble in a medium but which can be dispersed in it to modify its colour. Many water-soluble dyes can be converted into pigments by forming insoluble metallic salts or lakes, one of the earliest being the formation of the aluminium salt of cochineal mentioned in chapter 1. These dispersed particles are of a size between 100 and 1000 nm so that if the refractive index at any wavelength is substantially greater than that of the medium, light scattering will occur in addition to absorption and transmittance. This is the reason why there are white pigments but no white dyes – dyes on the fibre do not form aggregates nearly large enough to cause significant scattering. Light scattering by coloured pigments is sufficient to complicate the calculations involved in computer match prediction as we shall see in chapter 12 but it is normally quite insufficient to produce the required shades: a paint containing only Prussian blue, for example, would have to be sold as a black paint to avoid infringing trade description acts. Many coloured pigments have therefore to be mixed with about an equal amount of a white pigment to obtain attractive colours.

Pigments are usually applied by two quite separate methods:

(*a*) Surface coatings. This is, of course, the modern term for the traditional methods of painting, the pigment(s) being ground into the medium to make a paint or a printing ink which is then applied to the surface of the article being coloured.

(*b*) Mass-coloration. This method of application is used for colouring plastics before they are extruded. One of the earliest applications was to add the pigment to a solution of a man-made textile fibre such as viscose or cellulose acetate before spinning but by far the most important application today is to incorporate the pigment into molten polymer before extrusion. Often the polymer is coloured by a specialist company and extruded in ribbon form which is then cut into granules for re-melting and extruding into the shape of the final article by the manufacturer.

Pigments are also applied to textiles but as the method involves conventional textile machinery this application is best considered in the next section.

Dyes

With two exceptions which will be discussed later, dyes are applied to textiles, paper and leather from an aqueous solution or dispersion. In a dyeing operation the material is immersed in a dye-bath or padded with the dye liquor and in textile printing the dye is confined to specific areas by being applied as an aqueous paste. Once the dye has diffused into the fibre, some mechanism must restrict its return to the aqueous phase so that within a reasonable length of time the bulk of the dye has been transferred to the fibre. This is the fundamental mechanism involved in dyeing and when it occurs the dye is said to possess *substantivity* for the fibre in question.

There are six different mechanisms which confer substantivity, four chemical, and two physical.

Mordanting

This was the method widely used during the natural dye era and involved pre-treatment with metallic salts which subsequently formed additional linkages with dye molecules. Today the use of mordant dyes is mainly confined to wool using chromium salts which are invariably applied after the dye (after-chroming) or with the dye (the metachrome method). The CI generic name for this class of dye is *mordant*.

Salt linkage

Wool, silk and nylon contain amino groups which under acid conditions are ionised to give positively charged cations. Dyes containing sulphonic acid groups – anionic dyes – are negatively charged and, during dyeing, salt linkages are formed

$$\text{Fibre-NH}_3^+ + \text{Dye-SO}_3^- \rightarrow \text{Fibre-NH}_3^+\,\text{Dye SO}_3^-.$$

The CI generic name of these dyes is *acid* but if they possess substantivity for cellulose, as described next, they are classed as *direct*. Salt linkages can also be formed between the acidic groups in wool and silk and cationic dyes which constituted most of the earliest synthetic dyes such as magenta, malachite green and methylene blue. Such dyes are termed *basic* and whilst they were too fugitive for wide usage on natural textile fibres, they are much faster on the synthetic fibre, acrylic and constitute the major dyeing class for that fibre.

Hydrogen bonding

This is the mechanism whereby certain anionic dyes possess affinity for cellulose. The dyes are retained on the fibre by forming hydrogen bonds either with the fibre or between individual dye molecules to build up large aggregates which are physically trapped within the fibre pores. Such dyes are termed *direct* dyes and are characterised by having long planar molecules such as congo red (XXXIV) (CI Direct Red 28:22120) the first direct dye to be discovered (1884).

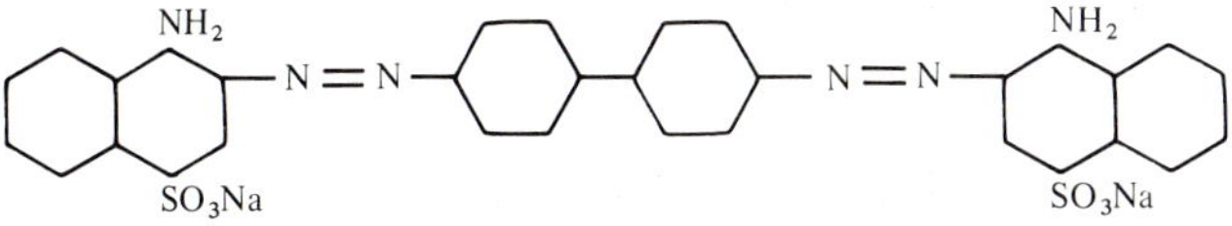

(XXXIV)

Covalent bonding

This is a comparatively recent discovery which again illustrates how dye chemists can fail to achieve their objective but instead discover something of very much greater importance. Stephen, of ICI, had synthesised azo dyes containing trichlorotriazine (XXXV) in the hope that one of the chloro groups would react with the amino groups in wool to form covalent links. These would be expected to be much stronger than the reversible salt links formed when wool is dyed with acid dyes and such dyes would therefore be expected to be much faster to washing. Rattee evaluated these speculative dyes but was not impressed with the results obtained and during the subsequent discussion in October 1953 Stephen commented that the reaction they were trying to achieve was the Schotten-Baumann reaction, familiar to generations of organic chemists. This is really only effective under strong alkaline conditions which would ruin wool.

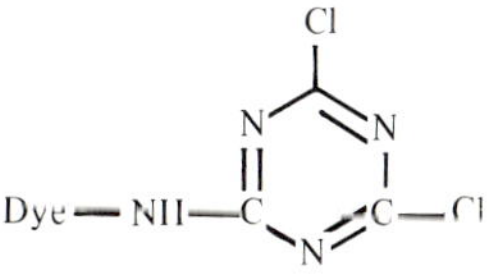

(XXXV)

The Schotten-Baumann reaction, however, also occurs with hydroxyl groups which are present in abundance in cellulose which is not adversely affected by high pH values. Rattee applied these speculative wool dyes to cotton and one of the most important classes of dyes was born, *reactive* dyes, which were first marketed in 1956 under the brand name Procion, exactly a century after Perkin's discovery of mauve. Monochlorotriazine dyes were also found to be just as valuable as reactive dyes: the manufacture of these had been patented many years earlier by Ciba but they had not realised their enormous potential. Ciba could, however, prevent ICI from *manufacturing* monochlorotriazine dyes whilst ICI could prevent Ciba from *selling* them for cellulosic fibres so, by mutual agreement, both firms entered the market with both mono- and di-chloro reactive dyes (ICI 1979). Rattee and Stephen were awarded the Perkin Medal of the Society of Dyers and Colourists for their invention.

Insolubilisation

This is the mechanism by which the insoluble colouring matter, indigo, was dyed. It was rendered soluble in alkaline solution by vatting and once inside the fibre it was oxidised back to the insoluble product, the microscopic particles of dye requiring aggregation by soaping into larger particles if maximum fastness was to be achieved. Indigoid and anthraquinoid vat dyes are classed as *vat* dyes and a similar mechanism is involved with *sulphur* dyes.

In 1880 the English dye maker, Read Holliday, successfully used the insolubilisation method to form insoluble azo dyes within cotton fibres. The cotton was first padded in an alkaline solution of β-naphthol followed by immersion in a bath containing a diazotised amine. This became one of the major methods of obtaining fast shades on cellulosic fibres, particularly red shades which did not feature prominently amongst the vat dyes. The components are listed in the *Colour Index* as CI Azoic Diazo Components and Azoic Coupling Components.

Other important dyeing methods are based on insolubilisation. In 1863, Lightfoot discovered a method of impregnating cotton with aniline and oxidising it *in situ* to form the fast aniline black (CI Oxidation Base 1:50440) and this technique with other bases became the standard method of dyeing natural furs.

The discovery of the pigment phthalocyanine blue in 1934 led to a search for ways of using it as a dye. Sulphonation gave a blue direct dye of good light fastness but very poor wet fastness (CI Direct Blue 86:74180) but in 1951 a method of solubilising it by chlormethylation followed by reaction with a tertiary base was discovered by ICI. During dyeing or printing, the solubilising groups were eliminated and an insoluble phthalocyanine derivative deposited within the fibre. Unlike the phthalocyanine blue pigment, this had a bright turquoise shade and was the first dye of that shade to have good all-round fastness properties. It had an effect on the world of fashion comparable with that of magenta and its brand name, Alcian Blue 8GX (CI Ingrain Blue 1:74240) passed into the vocabulary of designers. Another method of generating an insoluble phthalocyanine in the fibre was invented by Bayer in 1953. This was to print cloth with a paste containing di-iminoisoindoline (XXXVI) and a copper salt: on heating this formed copper phthalocyanine (CI Ingrain Blue 2:2 74160). It is very surprising that such a complex structure as (XX) is formed so readily from such a simple precursor.

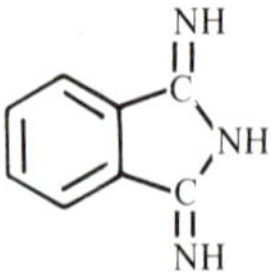

(XXXVI)

Solid solution

Certain dyes, insoluble in water and possessing substantivity for one or more hydrophobic fibres, are applied from aqueous dispersions. These dyes are termed *disperse* dyes and must be of small molecular size to achieve penetration into the fibre. It is often necessary to add fibre plasticising agents – wrongly termed 'carriers' – or alternatively to dye under pressure at temperatures well above the boiling point of water.

Two methods of applying disperse dyes to textile fibres without the use of water have been developed – solvent dyeing and transfer printing.

Solvent dyeing. The increasing cost of water and problems with effluents led to much effort being devoted to applying disperse dyes from solutions in solvents such as perchloroethylene instead of from aqueous dispersions. These have not, however, led to any widespread industrial application which is, perhaps, fortunate as no solvent recovery plant can ever be 100% effective and the dangers of atmospheric pollution by chlorinated hydrocarbons are well recognised.

Transfer printing. Many disperse dyes sublime and in 1968 a technique was perfected by the French textile company Filatures Prouvost-Masurel working in collaboration with Ciba and a paper-printing company whereby such dyes were first printed on to paper. Polyester fabric was then passed in contact with the printed papers through heated rollers which caused the dyes to sublime and dissolve in the fibre preserving faithfully the outline of the print. By 1973, transfer printing was being used for about one-fifth of all polyester fabric printing and its importance was recognised by the award of the Perkin Medal to de Plasse in 1977.

Pigment printing

Although pigments were being printed on textiles before 1937 using blood albumen, starch or glue as a binder the resultant fabrics had an unacceptable handle and only moderate wet fastness. In that year, however, the Interchemical Corporation introduced the *Aridye* system based on synthetic copolymers which gave a much softer handle and adequate fastness properties. Since then many new and better binders and thickening agents have been developed so that today pigment printing accounts for about one half of all textile printing.

The application of dyes to non-textile substrates

Dyes are used for colouring many materials other than textiles, the most important being paper, leather and foods. In paper manufacture the dyes are added to the pulp, i.e. a cold-dyeing operation, the dyes being conventional textile dyes whose properties and price are suitable for the industry. Hides and skins are dyed after tanning by tumbling them in drums containing dye solutions as hot as the leather will withstand. Many dyes have been specially synthesised for leather, particularly browns, as the conventional way of dyeing brown shades by combination of red, yellow and blue (or black) dyes is not satisfactory, the various components penetrating leather to different extents resulting in separation being apparent on cut edges.

The use of dyes for food coloration dates back to very early times, madder and kermes being used to colour the sugar imported into Europe from Alexandria from the twelfth century onward whilst cochineal was probably used even earlier. Inorganic pigments, including many lead compounds and copper arsenite were widely used for colouring sweets until the Adulteration of Food and Drink Act (1860) and the Public Health Act (1875) required local authorities to appoint public analysts and to institute regular sampling procedures in an attempt to detect adulterants 'injurious to health'. Fortunately many of the early synthetic dyes were technically superior for food coloration and few were believed to be injurious to health if free from contamination. Thus there was virtually no restriction on their use until a government committee recommended in 1924 that only colours whose harmless character had been demonstrated should be used. The government, however, adopted a much less restrictive policy: only five synthetic dyes were prohibited and it was not until 1957 that the UK followed the practice adopted in the majority of other developed countries of having a restricted list of permitted colours, all others being prohibited. Today the EEC permits 11 synthetic dyes and the USA 8. Whilst most of these permitted dyes are highly purified textile dyes, one or two have been specially synthesised for food coloration, the most important being Allura Red (CI Food Red 17:16035) which was developed in the USA in anticipation of the eventual prohibition of all other bright bluish reds. Allura Red, however, has neither UK or EEC approval as the alternatives have not been banned. For a recent, comprehensive review of developments in food colours the reader is referred to Walford (1980).

The fastness properties of dyes

The fact that dyes varied in their fastness properties had been recognised in the eleventh and twelfth centuries when dyers' guilds, separate from weavers' guilds, were formed in Europe. Dyers were divided into those capable of dyeing the faster dyeings – *grand teint* – and those only permitted to use less fast dyes – *petit teint*. Colbert, Controller General of Finances in France in 1661 is credited with being the first to specify a fastness test (boiling in slightly acid water) to differentiate between the two but it was du Fay de Cisternay, in 1737, who introduced a rational system which recognised the need to test separately for light fastness and for wet fastness. The manufacturers of synthetic dyes devised their own tests but it was not until 1914 that standard methods were introduced, the Verfahren, Normen und Typen of the German Fastness Commission. The British Society of Dyers and Colourists published a limited number of testing methods in 1934 followed by a much larger number in 1948 and after the Second World War the various authorities collaborated under the auspices of the European Colourfastness Establishment and the International Organisation for Standardisation (ISO) so that today the same testing methods are in use throughout the world. The complexity of the subject is well illustrated by the fact that the current International Standard 105/A contains 56 test methods (ISO 1982).

These are obtainable from the national standardising authority: in the UK this is the British Standards Institution† and the test methods are BS 1006: 1978.

The basic principle underlying all the tests except that for light fastness is to apply a laboratory test method designed to simulate the effects of the destructive agency such as washing to a specimen of the dyed textile in contact with specimens of several undyed fabrics of different kinds. The effect on the specimen is then assessed by comparing the perceived colour difference between the original and the treated specimen with a Grey Scale. This consists of pairs of cards painted grey illustrating colour differences increasing from zero (Grade 5) to a very large difference (Grade 1). The staining of each adjacent fabric is similarly assessed against another scale in which the illustrated differences are set against a reference white card (Fastness Tests Co-ordinating Committee 1953).

The determination of light fastness is inevitably different because there is no typical dosage of radiation to which the textile will be exposed during its lifetime: a gown will rarely be exposed to daylight, a curtain may be hung in a south-facing window for several years, whilst an awning will have to withstand repeated wetting as well. The method used in the majority of countries is an elaboration of that first developed by du Fay de Cisternay. A specimen of the textile under test is exposed to daylight or to arc-light together with eight standards – pieces of cloth each dyed with a different dye whose resistance to exposure varied from very low to very high. The dyes were selected following a comprehensive investigation which showed that the fading rates of dyes on wool were least sensitive to variations in humidity and that blue dyes would provide a set nearest to the ideal spacing which was that each dyeing would be twice as resistant as the one below it (Cunliffe 1930). Standard I shows detectable fading after only a few hours exposure to bright sunshine whereas Standard 8 would require at least two summer months exposure in southern England. The specimen and standards are exposed for two periods to produce two degrees of fading of the specimen which are defined by the Grey Scale already described (Grades 4 and 3) and the light fastness of the specimen is the number of the standard which fades at the same rate.

This method for light-fastness determination is not used in the USA. In that country, fading lamps have always been favoured over natural daylight and these are calibrated using reference fabrics. The most common method of specifying light fastness is to give the number of Standard Fading Hours required to produce a colour difference visually equal in magnitude to the pair defining Grade 4 on the Grey Scale.

Many of the ISO fastness testing methods are officially specified for leather and are also widely used for other substrates such as paper: the light-fastness testing method is also widely used for paints and plastics.

† 2 Park Street, London W1A 2BS; 3 York Street, Manchester M2 2AT.

Standard depths

Strictly speaking it is incorrect to refer to the fastness property of a *dye* as this characteristic only applies to a dyed substrate. On any substrate most fastness properties depend on the depth of shade and to ensure that any comparisons made between the data published by different dye manufacturers were valid the German Fastness Commission in the 1930s established a series of dyeings in 18 hues of the same depth, termed 1/1 Richttyptiefe (Standard Depth). These were chosen visually but no details are available as to how the choice was made (Kuehni 1978). In 1954 they became an ISO Standard and were augmented by dyeings illustrating 1/25, 1/12, 1/6, 1/3 and 2/1 Standard Depths. These were obtained by dyeing wool to complete exhaustion with level dyeing acid dyes with the specified fractions and multiple of the amount required for the 1/1 Standard Depth. They were also matched on lustrous viscose pieces but subsequently, in the UK, both were replaced by matt-painted cards (BS 1006: Section A01 SDM). As the standard depths are illustrated in only 18 hues, the selection of the 1/1 Standard Depth from a series of dyeings of markedly different shades is very difficult. This became a major problem following the introduction of reactive dyes in 1956 as many were very much brighter than those used to dye the standard depths. The obvious solution is to define standard depths colorimetrically and methods for doing this have been published (Rabe and Koch 1957, Gall and Riedel 1965). Although the establishment of such a method has been on the working programme of ISO/TC38 since 1981 neither of these methods has been proposed which seems to confirm Brockes's view that neither is really suitable (Brockes 1975).

The evaluation of cost effectiveness

Textiles are normally coloured with a mixture of three dyes and the number of three-dye combinations in each dyeing class which will be technically suitable for most shades is very large. We shall see in chapter 12 that the greatest benefit of computer match prediction is the identification of the cheapest combinations which will match a given shade. The determination of the overall cost effectiveness of a dye for most of the shades in which it could be a major component cannot readily be made by this method but Derbyshire and Marshall devised another which has been very successful.

When increasing amounts of any dye are applied to a substrate the colorimetric value which increases approximately linearly is the K/S value which is the basis of computer match prediction. At each wavelength the K/S value is multiplied by the relative spectral power distribution of the illuminant and by the sum of the colour matching functions at that wavelength. These products are then summed to give what they term the *Integ* value (Derbyshire and Marshall 1980). There is, of course, no psychophysical justification for this calculation and it by no means quantifies equal strengths for dyes of different hues. The *Integ* values of the ISO Standard Depths have a coefficient of variation from the mean of 24% (Kuehni 1980). The cost effectiveness of all dyes of the same general colour, for example all brown dyes, is however reliably given by the cost of achieving the same *Integ* value.

Factors contributing to the growth of synthetic colouring matters

It was mentioned in chapter 1 that about 10 000 separate synthetic organic compounds had been marketed as dyes or pigments since Perkin's mauve, the annual rates at which new dyes have been introduced being displayed in figure 1.1. About 4000 of these are no longer manufactured anywhere in the world, about 800 are pigments and over 350 are fluorescent brightening agents which means that no fewer than 4800 dyes are available for dyeing textiles, leather, paper and food. This large number is in no way attributable to the fact that there are at least a million colours which are perceptibly different. There are in fact several factors, both technical and commercial, contributing to the large number of available dyes. The chief factors are listed below.

(*a*) Each dye possesses substantivity for only a limited number of fibres.

(*b*) Each fibre can be dyed with dyes of more than one dyeing class. For example, cellulosic fibres – by far the most important in terms of weight dyed per annum – can be dyed with the following classes of dye: direct; indigoid vat; anthraquinonoid vat; sulphur; solubilised vat; azoic and reactive. Each class has certain overall characteristics which make it particularly attractive for certain end uses: dyes in the direct class are cheapest and the easiest to apply; reactive dyes give the brightest shades; vat and azoic have maximum fastness.

(*c*) Within each dye class, individual members vary widely in fastness properties, not only in general but specifically: a dye of high light fastness may have poor washing fastness and vice versa.

(*d*) A factor which is at least as important as all of these is the competitive nature of the dyestuff manufacturing and using industries in the capitalist countries. This is perhaps best illustrated by what happened in 1956 after ICI and Ciba launched the first reactive dyes, the mono- and di-chloro triazinyl dyes. Intensive research activity was started by their competitors to discover other groups conferring fibre-reactivity which had not been patented. Many attempts were successful so that today there are no fewer than fourteen chemically different classes of reactive dyes and the total number marketed is over 700.

4

The Main Theories of Colour Vision

The Ancient Greek philosophers were fascinated by colour and put forward several theories, some of which were quite bizarre. Plato, for example, suggested that an inner fire gave rise to visual rays shooting outward from the eye towards an object. Epicurus, however, completely rejected this emanation theory and suggested that coloured corpuscles from the object entered the eye. They were unanimous, however, in believing that light when pure, was white and this belief was universally accepted until Newton's classic experiments led to the growth of colour science. Newton was not, however, interested in developing theories and it was not until 1777 that a substantial theory was developed by George Palmer the essence of which was that all colours of light are actually composed of three primary colours only, red, yellow and blue and that 'the surface of the retina must be composed of three kinds of fibres, or membranes each analogous to one of the three primary rays and susceptible to being stimulated by it'. This was published in a book entitled *Theory of Colours and Vision* but it had no impact whatsoever in the development of colour science and its very existence only emerged when Gordon Walls was searching for the earliest reference to defective colour vision: his account of this makes delightful reading (Walls 1956). An identical view was subsequently put forward by Thomas Young, undoubtedly in complete ignorance of Palmer's original suggestions, and this formed the basis of the trichromatic theory of colour vision whose importance to the development of colour science cannot be exaggerated.

The trichromatic theory

Young propounded his theory in the Bakerian Lecture he gave to the Royal Society in 1801. This lecture was primarily concerned with developing the wave theory of light but he concluded it with the following (Young 1802a):

> As it is almost impossible to conceive each sensitive point of the retina to contain an infinite number of particles, each capable of vibrating in perfect unison with every possible undulation, it becomes necessary to suppose the number limited, for instance to the three principal colours, red, yellow and blue, and that each of the particles is capable of being put in motion more or less forcibly by undulations differing less or more from perfect

> unison. Each sensitive filament of the nerve may consist of three portions, one for each principal colour.

The concept that there are three principal or primary colours and that these are red, yellow and blue was already well established. Robert Boyle, writing in 1664, mentioned that these colours, together with white and black, were those used by painters. This concept is sound: the number of different colours obtainable by mixing red, yellow and blue paints is much greater than if any other three-colour combinations are used and the increase obtainable by having four or more is not particularly great.

Although there were many studies made and published on the effects of mixing pigments, very few experiments were carried out by mixing coloured lights and all scientists tacitly assumed that the effects would be similar. But seven months after his Bakerian Lecture Young modified his choice to red, green and violet 'in consequence of Dr Wollaston's correction of the prismatic spectrum' (Young 1802b). Now these happen to be the colours of the three lights yielding the greatest number of other colours in mixtures and the recognition of this fact was a step no less important to the development of colour science than Newton's discovery of the true nature of white light. But, strangely enough, this did not occur until fifty years later: Young's modification resulted from a completely false deduction.

Wollaston (1802) had achieved much greater prismatic dispersion than Newton and had observed seven dark lines in the spectrum. He thought that five of these were physical boundaries of the principal spectral colours and said that these were not seven, as Newton had claimed, but four – red, yellowish-green, blue and violet. Young repeated this experiment 'with perfect success' and modified his choice of primaries without actually specifying how he chose these three from Wollaston's four. This interpretation of the dark lines was, of course, completely wrong. In 1817 Fraunhofer achieved even greater prismatic dispersion, counted 325 lines and showed that they were caused by certain wavelengths of light being absorbed by elements in the solar atmosphere. Young also contributed another error: he said that 'within their respective limits the colours differ scarcely at all in quality' (Young 1807). Newton had, however, been absolutely right when he said that the spectrum contained 'an indefinite variety of intermediate gradations' in addition to the well-known seven colours.

Young realised that if his theory were correct a combination of red, green and violet colours must cause a sensation of white. The method he used to investigate this was to paint red, green and violet sectors on a disc and to spin it rapidly enough to eliminate flicker. The best one can achieve by this method, is however, a neutral grey. Young also claimed as proof the following (Young 1807):

> If we mix together, in proportions, any substances exhibiting these colours in their greatest purity, and place the mixture in a light sufficiently strong, we obtain the appearance of perfect whiteness.

It is obvious that this statement must have owed far more to wishful thinking than to visual observation as any such mixture must appear as a dark grey under any normal viewing conditions irrespective of the light intensity. The significant fact is, however, that it shows that Young was also one of the scientists who had not realised the fundamental difference between mixing lights and mixing pigments because the first experiment involved the former and the second the latter. The fact that from completely fallacious reasoning he made the correct choice must be regarded as astonishingly good luck.

Helmholtz (1852) was the first to recognise that there was a fundamental difference between mixing lights and pigments which could be explained by the established laws of physics. It can best be illustrated by considering one of the most important colour combinations – mixing blue and yellow dyes or pigments to produce green colours – though it is amusing to note that Aristotle and Leonardo da Vinci claimed that green was a colour which could not be obtained by mixing. It has already been mentioned that scientists assumed that the empirical laws of colour mixing based on the behaviour of pigments applied equally to mixing lights and this belief has even persisted up to the present day with a training manual for technicians stating that a mixture of yellow and blue *lights* appears green. The simple experiment of mixing yellow and blue lights, however, clearly shows that the only way such a mixture can appear even slightly greenish is to use either a greenish yellow or a greenish blue: pure yellow and pure blue lights when mixed in the correct proportion appear white.

If we consider the spectrum to consist of red, yellow, green, blue and violet (*R*, *Y*, *G*, *B* and *V*) lights then the mechanism put forward by Helmholtz to explain the colour perceived when yellow and blue pigments are mixed together is as follows:

The yellow pigment scatters *RYG* and absorbs *BV*.

The blue pigment absorbs *RY* and scatters *GBV*.

When the *RY* and *G* lights scattered by the yellow pigment fall on a particle of the blue pigment, *R* and *Y* are absorbed, *G* is scattered.

When the *GB* and *V* lights scattered by the blue pigment fall on a particle of yellow pigment, *B* and *V* are absorbed, *G* is scattered.

The only light which can enter the eye is that which is not absorbed by either pigment but scattered by both: the colour of this light is green.

The behaviour of dyes is similar but the light which is not absorbed is not scattered by the dye but is transmitted to be scattered by the substrate. Because the effects of mixing dyes and pigments are governed by their power to subtract certain regions of the spectrum from the incident light, this is called *subtractive colour mixing* in contrast to direct light mixing which is termed *additive colour mixing.*

The ideal subtractive primaries

The most important practical consequence of this differentiation was the discovery that the *ideal* subtractive primaries were not actually red, yellow and blue. Helmholtz had shown that the ideal additive primaries were monochromatic red, green and violet lights but a close enough approximation to these can be obtained by dividing the spectrum into three portions, the dividing lines being in the yellow and blue-green regions, and recombining the components in each portion. The colours of the resulting lights are red, green and blue and mixed in the correct proportions virtually all the colours which can be obtained with dyes and pigments can be matched. As dyes and pigments achieve their colour effects by absorption it follows that if three are chosen which absorb only the red, the green and the blue spectral regions respectively these will also yield the maximum number of colours. The colours of these ideal primaries can be deduced as follows.

(*a*) The colorant absorbing the red spectral region will absorb neither the blue nor green regions and will therefore be blue-green in colour.

(*b*) The colorant absorbing the green region will absorb neither the red nor blue regions and will therefore be purple in colour.

(*c*) The colorant absorbing the blue region will absorb neither the red nor green regions and will therefore be yellow in colour.

Existing blue-green and purple colorants nearest to the hues of the ideal subtractive primaries are, however, unsatisfactory, because they are too dull. In practice, therefore, a bright greenish-blue and a bright bluish-red are used. The former colour is termed cyan, from the Latin *cyaneus*, the latter, magenta, whose origin was described on page 14. Their effects in mixtures are often surprising and unpredictable by those familiar with the behaviour of conventional red, yellow and blue primaries. They can best be remembered by realising that the method of derivation means that pairs of additive primaries generate the subtractive primaries and that pairs of subtractive primaries generate the additive.

Additive	*Subtractive*
red + green → yellow	yellow + cyan → green
green + blue → cyan	cyan + magenta → blue
blue + red → magenta	magenta + yellow → red

A mixture of three additive primaries, of course, gives white whilst a mixture of the three subtractive primaries gives black.

The discovery that the best subtractive primaries are not red, yellow and blue but magenta, yellow and cyan had no effect whatsoever on the choice of dyes and pigments used in the dyeing and paint industries or in the colouring of plastics. The reasons for this were two-fold. Firstly there are very few magenta and cyan colorants of adequate fastness properties and secondly those that do exist absorb quite strongly in the regions where, ideally, no absorption should occur. The resultant colours produced are in consequence, somewhat dull and they are only used in those industries where the number of colorants has to be limited to three – the colour printing of non-textile substrates and colour photography. Conventional paper printing augments the magenta, yellow and cyan printing inks with a black as the best mixture gives a very poor black and if quality colour printing is required then as many as seven process inks plus black may be used. Unwanted absorption in colour photography has been overcome by a very sophisticated technique known as 'integral masking' which is described in Hunt's comprehensive book (Hunt 1967).

The laws governing additive mixing are fundamental in the sense that they describe the way the eye behaves. As long as it was assumed that these were the same as those laws believed to be governing subtractive mixing no further progress was possible. As soon as Helmholtz had recognised the difference between additive and subtractive mixing colour science began to develop very rapidly.

Hermann Günther Grassmann was a mathematician who devised a theoretical framework for colour perception. This suggested that Helmholtz had been wrong in claiming that only indigo and yellow were complementary colours (Grassmann 1853, 1854); Helmholtz subsequently confirmed experimentally that Grassmann was right (Helmholtz 1855). Grassmann also deduced that lights of the same colour but of different spectral composition behaved identically in mixtures. This led to the concept of colour matching equations which behave like algebraic equations and are the basis of modern colorimetry. Grassmann's achievements are all the more surprising when it is realised that he probably never carried out even elementary experiments on colour mixing: this is shown by the fact that he believed that the colour of the extreme red of the spectrum was the same as that of the extreme violet. This led him to conclude, erroneously, that every spectral colour has a complementary spectral colour whereas this is not true of the green region between 490 and 570 nm.

The quantification of colour matching equations was then carried out by James Clerk Maxwell who realised that this would permit a wholly objective method of colour measurement to be devised. Maxwell's method involved determining the amounts of three standardised primary lights required to match each part of the spectrum making further visual matching unnecessary. No other objective method is conceivable and Maxwell's method, which is widely used today, is the subject of chapter 8.

All of these studies were based on the belief that the trichromatic theory was correct and they yielded no evidence that this was not the case. There was also no simpler theory which could account for these facts and as these are the criteria for acceptability in the scientific community, this was the only theory taught in schools of colour chemistry and dyeing until recently. Largely as a result of the influence of Helmholtz's *Physiological Optics* (1866, translation 1924) it is generally known as the Young-Helmholtz trichromatic theory but Sherman, whose fascinating account of this period provided much of the matter referred to so far in this chapter, concluded that it would be fairer to call it the Young-Helmholtz-Maxwell theory (Sherman 1981).

Opposition to the trichromatic theory

In spite of the overwhelming, though indirect, evidence in favour of the trichromatic theory it was by no means universally accepted. There were two reasons for this. The first was that it was believed to be based on a false premise because (Guild 1932):

> Many physicists for some obscure psychological reason persist in stating it (the trichromatic theory) in a form which, as they are perfectly well aware, is grossly inaccurate, as when they state that any stimulus, whatever its physical characteristics (i.e. spectral energy distribution) can be matched, as regards the colour it evokes, by a mixture of three other stimuli.

This regrettable practice still occurs as the following quotations show:

> . . . in order to reproduce a colour, it is necessary to take only three radiations, such as red, a green and a blue, and mix them in the correct proportions [1964];

> The human observer can match a light of any wavelength by a suitable mixture of three 'primary' coloured lights [1977];

> In addition to the clear establishment of the proposition that all colours can be created by the combination of three spectral stimuli, Maxwell . . . projected a first trichromatic photograph [1979].

Helmholtz had originally rejected the theory for this very reason:

> If the sensation of yellow by the yellow rays of the spectrum were due to the fact that by them the sensations of red and green were simultaneously excited, and both working together produced yellow, exactly the same sensation must be excited by the simultaneous action of the red and green rays; nevertheless by the latter we can never obtain so bright and vivid a yellow as that produced by the yellow rays.

He said that it would only be possible to match the spectral colours if the number of primaries was five – red, yellow, green, blue and violet (Helmholtz 1852) but later he realised that this failure could be explained if each of the three cone types (types of retinal receptor) was sensitive to the whole of the spectrum (Helmholtz 1866, 1924). The spectral sensitivity curves he deduced were broadly similar to those of figure 4.1 which are the most recent (Estévez 1979, quoted by Hunt 1982).

The 'obscure psychological reason' envisaged by Guild is probably nothing more than carelessness in wording. Maxwell had demonstrated that *every* light, including the spectral colours, could be a component of a colour matching equation with red, green and blue primaries providing it was permitted to add one of the primaries to the light in question. The algebraic nature of colour matching equations meant that this amount could be treated as a negative

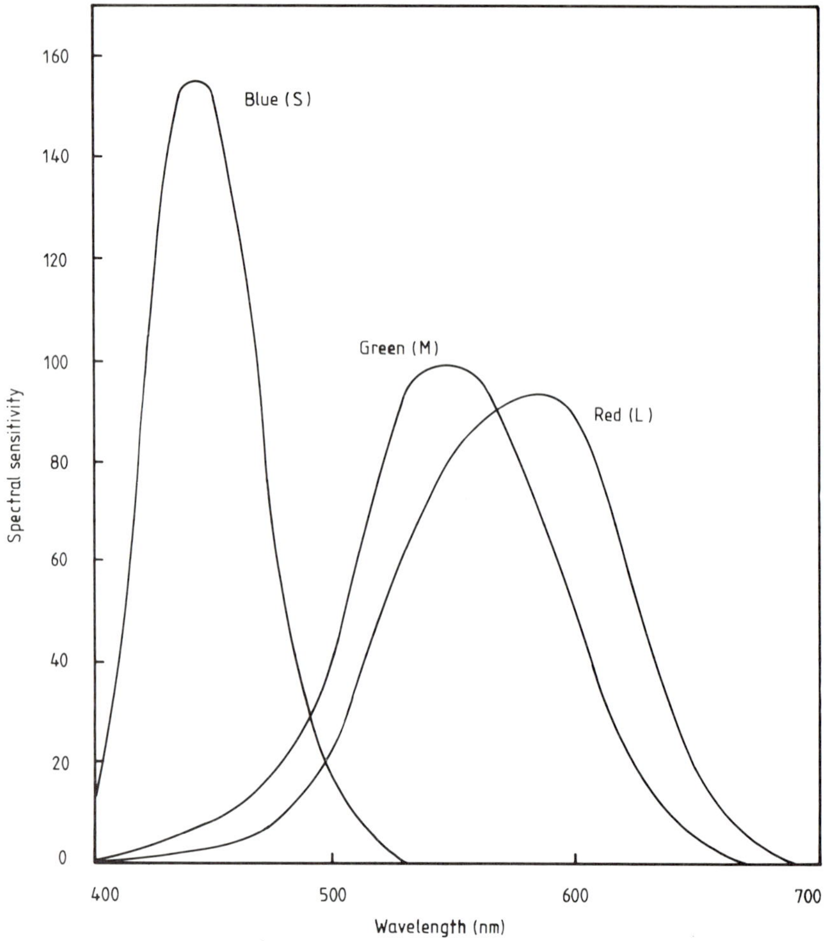

Figure 4.1. Cone sensitivity curves. S, short wavelength; M, medium wavelength; L, long wavelength.

number, for example, if

$$Q + R \equiv G + B$$

then

$$Q \equiv G + B - R.$$

To many physical scientists there is no fundamental difference between $Q \equiv G + B - R$ and $Q' \equiv G + B + R$ so they overlooked the fact that in the first instance they are not matching the colour of Q but an entirely different colour, $Q + R$.

The second, and by far the most important reason why many rejected the trichromatic theory was the belief Maxwell clearly expressed in his first paper on colour (Maxwell 1856):

> The theory which I adopt assumes the existence of three elementary sensations, by the combination of which all the actual sensations of colour are produced . . . .

An elementary colour sensation is one which cannot be described in terms of other sensations. For example the sensation described as purple can be adequately described to a person whose native language does not contain a word for purple as a very reddish blue or a very bluish red just as the English manage without a common word for blue-green. Purple is therefore not an elementary colour sensation. But it is impossible to choose *three* colour sensations which can be used to describe *all* colour sensations: if red, green and blue are chosen, the sensation of yellowness cannot be described and if red, yellow and blue are chosen that of green-ness cannot be described. The latter statement may cause surprise because many people, including Goethe and Brewster, genuinely believed that they could perceive both yellowness and blueness in a green: but this is only because of the familiarity with mixing paints. Any lingering doubts over this will be dispelled when it is realised that one can select a green shade which is *neither* yellowish *nor* bluish.

These four fundamental colour sensations are often called unique or unitary hues though a more descriptive term is *psychological primaries*. These are

(*a*) a red which is neither bluish nor yellowish: all spectral reds appear slightly yellowish;
(*b*) a yellow which is neither reddish nor greenish;
(*c*) a green which is neither bluish nor yellowish;
(*d*) a blue which is neither reddish nor greenish.

Dimmick and Hubbard determined the wavelengths of yellow, green and blue unique hues using ten observers. The median values were 582, 515 and 476 nm respectively. The wavelength of the light which was complementary to that of unique red was 493.6 nm (Dimmick and Hubbard 1939). It is significant that as a language develops these are the first hue terms to be devised, usually in that order though sometimes a word for green precedes that for yellow (Berlin and Kay 1969).

Dye and pigment technologists recognise the existence of four psychological primaries as they need four terms to describe hue differences. This fact was often reconciled with the trichromatic theory by claiming that yellowness and green-ness are merely different words to describe what is essentially the same effect. Support for this view was supposed to be forthcoming from the dye-makers' suffixes G, 2G, 3G etc which indicate both *greenish* blues and *yellowish* reds. This idea is completely fallacious however as it arises from the fact that this practice originated in Germany where the word for yellow (gelb) and green (grün) have the same initial.

Another similar fallacy occurred in an authoritative account of defective colour vision. It was correctly stated that the discriminations of a normal observer can be classified as light–dark, yellow–blue and red–green but this was then described as 'a corollary of trichromatism' which it certainly is not.

There is an important fact about these psychological primaries which is that not all of the possible combinations of pairs can be perceived simultaneously

Combined sensations

Possible	*Impossible*
redness + yellowness	redness + green-ness
yellowness + green-ness	yellowness + blueness
green-ness + blueness	
blueness + redness	

These facts are indisputable and clearly any theory of colour vision which does not explain them is not necessarily wrong but is certainly incomplete. The experimental psychologist Ewald Hering, in 1872–4, therefore developed another theory which did explain these facts – the opponent-colours theory (Hering 1878, 1964).

The opponent-colours theory

Hering postulated that there were three processes involved in the perception of colour, each process involving activity (katabolism) or repair (anabolism). One of these processes resulted in the sensation of redness if there was an excess of activity over repair, green-ness if the balance was in the other direction and neither if anabolism and katabolism were balanced. A second process resulted in the sensation of yellowness or blueness or neither whilst the third resulted in a sensation of white or black or, when balanced, grey. This theory does explain why there are four psychological primaries and why two pairs of hue sensations, redness and green-ness, yellowness and blueness cannot be perceived simultaneously whereas each member of one pair could be simultaneously perceived with either member of the other. The extension to black and white was not particularly convincing because most observers would regard grey as a combined sensation of blackness and whiteness rather than the presence of neither.

This theory was completely rejected by the physical scientists, but it found favour, however, among those artists who were interested in theories of colour vision and with experimental psychologists because it was much better at explaining colour perception. In this respect it was of greater value than the trichromatic theory in explaining the sensations of colourists but as colour technology developed from the physical sciences it was never taught until comparatively recently. This prevalent attitude is well illustrated by the fact that Sherman (1981) devoted a whole chapter to the physicist Brewster's bizarre theory of the spectrum (Brewster 1834) which was completely disproved by Helmholtz (1852), but did not even mention Hering. The rivalry between the supporters of the two theories was comparable to that between the supporters of the wave and corpuscular theories of light and heightened by an element of the 'two-cultures' syndrome. The trichromatic theory possessed the overriding advantage of the stimulus quantification provided by Helmholtz and Maxwell and it was not until 1955 that similar quantification was obtained for the opponent-colours theory (Hurvich and Jameson 1955).

The retinex theory

In 1964 Land put forward a new theory of colour vision involving interactions between cone signals which occurred either in the retina or in the cortex (hence the name). As this theory was developed to explain the remarkable characteristic of *colour constancy* it will be discussed in chapter 11.

The identification of three cone types

One fact which would have virtually proved the correctness of the trichromatic theory was the identification of three separate kinds of receptor in the retina. Two kinds had long been established, named after their shapes as rods, of which there are about 120 million, and cones, of which there are about 7 million. The rods are only involved at very low levels of illumination such as the light from a quarter-moon. At these light levels there are no colour sensations – 'at night all cats are grey'. But the recognition of three different kinds of cone did not occur until 1964 when measurements were made on single cones by microprojecting a tiny spot of light from a monochromator through them and into a spectrophotometer (Marks *et al* 1964, Brown and Wald 1964). The cones fell into three classes with absorption maxima at about 445, 535 and 575 nm. These measurements were a remarkable achievement in view of the minuteness of the cones – there are about 150 000 per mm^2 in the central fovea, the retinal region devoid of rods – the low pigment density and the need to use very weak lights to avoid pigment bleaching. Although many measurements on animal photoreceptors were subsequently made following advances in the design of microspectrophotometers the next measurements on rods and cones were not carried out until 1979 (Bowmaker and Dartnall 1980). Subsequent measurements on the retinas of eyes from seven persons (which were removed because of the presence of a melanoma) gave the results shown in table 4.1 (Dartnall *et al* 1983).

Table 4.1 Absorption peaks of rods and cones from seven persons.

Photoreceptor	Number	λ_{max} (nm)
Rod	39	496.3 ± 2.3
Red cone	58	558.4 ± 5.2
Green cone	45	530.8 ± 3.5
Blue cone	5	419.0 ± 3.6

Another method of cone differentiation is retinal densitometry developed by Rushton and his colleagues. A beam of light shone into the eye will be in part reflected back by the *choroid*, a pigmented layer at the back of the eye. If weak beams of light of various wavelengths are sent through the fovea absorption by the pigments in the cones occurs so that the composition of the emergent light will be different from that of the incident light. By first exposing the retina to intense red light the cones containing the pigment whose absorption peak is at the longest wavelength will be preferentially bleached and by measuring the absorption curve intermittently during regeneration, that due to these cones can be determined. Exposure to blue-green light followed by measurement gives the absorption curve of the pigment with the next lower λ_{max}. The curves they obtained after having built six different densitometers of increasing sensitivity were negligibly different from those determined by microspectrophotometry. They were, however, unable to apply this technique to the cones containing the blue-sensitive pigment as these are too few in number (Rushton 1975).

A third method of identifying the cone mechanisms in man was developed by Stiles utilising selective chromatic adaptation (see chapter 11 for a discussion on chromatic adaptation) to desensitise one cone mechanism more than the other two. By determining increment threshold functions the action of the lesser adapted mechanism could be ascertained. For an introductory essay and selected papers on this somewhat complicated but impressive approach see Stiles (1978).

From these and other studies, particularly on observers lacking one or other of the cone pigments (see chapter 6) Estévez derived the cone sensitivity curves shown in figure 4.1 (Estévez 1979, quoted by Hunt 1982).

Cone nomenclature

The three cone types have long been called red, green and blue and it is often assumed that this refers to the colour of the monochromatic light causing maximum excitation. This is not so in the case of the red cones as the peak of the excitation curve lies in the greenish-yellow region of the spectrum. Another assumption is that suggested by Helmholtz: excitation of only one or other of the three cone types would result in one or other of these hue sensations. This is also fallacious. Light at the short wavelength end of the spectrum only excites the blue cones but although the predominant sensation is that of blueness this is accompanied by a degree of redness sufficient to justify a different hue term, i.e. violet. At the other end of the spectrum an increase in wavelength increases the ratio of red to green cone excitation but the sensation of yellowness which accompanies that of redness does not continually diminish. Beyond 700 nm the hue does not change and light of these wavelengths causes a sensation of yellowish-red.

These anomalies have resulted in several attempts being made to persuade authors and teachers to abandon the terms red, green and blue in favour of long, medium and short, α, β and λ or by 575, 535 and 445 (λ_{max}) but without success. In 1985 the most commonly used terminology was still red, green and blue and this will be used subsequently, usually abbreviated to R, G and B.

The identification of opponent-colour mechanisms

The final proof of the existence of three cone types did not mean the abandonment of the opponent-colours theory, however, because at about the same time direct proof of the correctness of the fundamentals of the theory was also forthcoming. How this came about is best left until the next chapter which will deal with the mechanisms of colour vision as presently understood.

5

The Mechanisms of Colour Vision

The chemical constitution of the photoreceptors

Night blindness – insensitivity to dim light – is one of the oldest diseases known to man and Ancient Egyptian medical papyri contain the recommendation that it can be cured by a diet of liver. In Denmark during the First World War night blindness was shown to be a symptom of a deficiency of vitamin A, the concept of vitamins having been put forward in 1911, and one of the richest sources of vitamin A was liver. At that time the ways in which the vitamins fulfilled their role of being vital for normal health and growth were completely unknown but no one suspected them of being directly involved in physiological processes. The structure of vitamin A (XXXVII) was established in 1931. The main source of the vitamin is the cleavage in the body of β-carotene (XVIII).

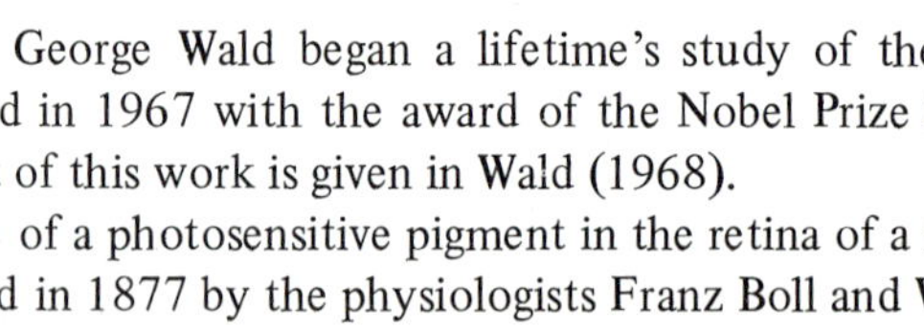

(XXXVII)

At this time George Wald began a lifetime's study of the visual pigments which culminated in 1967 with the award of the Nobel Prize for Physiology. A personal account of this work is given in Wald (1968).

The existence of a photosensitive pigment in the retina of a frog was independently discovered in 1877 by the physiologists Franz Boll and Wilhelm Friedrich Kühne. Boll noticed that when a frog's retina was removed in the dark it appeared bright red but on exposure to light it became yellow and finally colourless. Kühne found that this also occurred with many other animals: in the living animal the colour is restored in the dark but not in excised retinas. Wald found that this bleaching process generated vitamin A via an intermediate compound he termed 'retinene' which had formed the visual pigment by combining with a protein, 'opsin'. Morton and his co-workers (Ball *et al* 1946) then showed that retinene was the aldehyde of vitamin A and that combination with opsin was via the well-known Schiff's base reaction

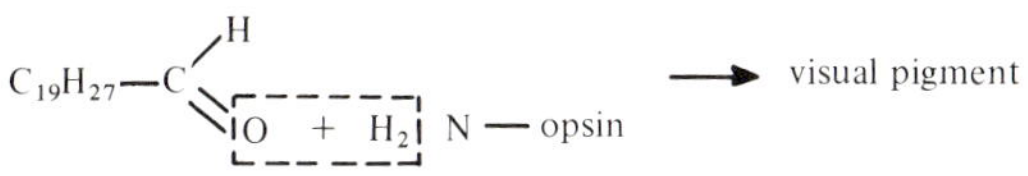

the amino group being later found to be the ϵ—NH_2 group of lysine (Bownds 1967). This is, literally, the only simple reaction involved in the visual process.

In 1960 the International Union of Pure and Applied Chemistry recommended that vitamin A should be re-named retinol and retinene, retinal. At the same time the carbon numbering system in (XXXVII) was introduced, each carbon-carbon bond being identified by the lower number.

Hundreds of visual pigments have since been discovered and they are all based on retinal or on dehydroretinal which contains two fewer hydrogen atoms and a double bond at position 3. When different visual pigments were first identified they were given trivial names such as rhodopsin for the human rod pigment. As the numbers increased, however, this method became unworkable and as it had been found that adequate identification was given by the retinal involved and the wavelength of maximum absorption, λ_{max}, this method was adopted instead. All pigments based on retinal are now termed rhodopsins whilst those based on 3-dehydroretinal are termed porphyropsins, the name given by Wald to the first such pigment identified in freshwater fishes. As the differences between the chromophores are usually insignificant, the term retinal should be taken to include both unless otherwise stated: similarly the term rhodopsin should be taken to include porphyropsins as well.

The structure of rhodopsin

Retinal exists in several cis-trans isomers but all the naturally occurring visual pigments are based on the 11-cis isomer: this is not only bent at the cis linkage but twisted as well. The only other isomer with a similar configuration is the synthetic 9-cis and as this will also combine with opsins to form photosensitive pigments, isorhodopsins, Hubbard and Kropf (1958) suggested that the opsin contained an accurately formed niche into which these and no other isomers would fit.

The λ_{max} values of all the retinals cover the range 368–381 nm and mere Schiff's base formation with an opsin causes a slight blue-shift but when the 11-cis (or the 9-cis) isomer combines with an opsin there is usually a marked red-shift, λ_{max} values for rhodopsins ranging from 430–569 nm and those for porphyropsins from 510–625 nm. There is one intriguing exception; the lacewing *Ascalaphus macaronius* is sensitive to ultraviolet light, its visual pigment is a retinal–opsin complex but λ_{max} is *blue-shifted* to 345 nm (Gogala *et al* 1970).

The porphyropsin formed from a given opsin always has a higher λ_{max} than that of the corresponding rhodopsin and this is utilised in a most ingenious way by amphibia, a discovery originally made by Wald in 1945 (Wald 1945). The tadpole of *Rana pipiens*, for example, has three receptors based on porphyropsin with λ_{max} values of 438, 523 and 620 nm, the 3-dehydroretinal being produced by the animal as there are no corresponding carotenoid pigments occurring in nature. When it metamorphoses, however, this production ceases and the opsins then combine with retinal to form rhodopsins with λ_{max} values of 432, 502 and 575 nm. Receptors with maximum sensitivity to longer wavelengths are better suited to animals living in an aquatic environment as such environments contain blue-absorbing and scattering materials: those at shorter wavelengths are better suited to land-based animals.

The red-shift occurring when 11-cis retinal combines with an opsin is a most unusual occurrence as nothing comparable occurs between dyes and proteins. Two facts must be significant. The first is that if rhodopsin absorption curves are plotted on a frequency basis rather than the customary wavelength scale, their shapes are virtually identical (Dartnall 1953). The same is true of the porphyropsins though the shape is different (Bridges 1967, Munz and Schwanzara 1967). In 1982 it was found that curve shapes for monkey receptors were more similar if plotted with the abscissa representing the fourth root of wavelength (Barlow 1982) and this was confirmed for human receptors (Dartnall *et al* 1983). The second is that the λ_{max} values of rhodopsins are not randomly distributed throughout the relevant wavelength range but exhibit pronounced spectral clustering around certain wavelengths (Dartnall and Lythgoe 1965, Bridges 1965). One theory put forward to explain these facts is that of Dartnall and Lythgoe. They postulated that a pair of dipoles which might exist only transitorily in the conjugated chain of the retinal portion of the molecule may be stabilised by electrostatic attraction to oppositely charged atoms in the opsin portion. They consider that twenty pairs of dipoles may be involved, each resulting in a different λ_{max}. The most recent theory is that of Honig and Dinur who explain the red-shifts by there being interactions between one negatively charged amino acid side chain in the opsin and the C12 and C14 carbon atoms of the retinal and between another side chain and the protonated Schiff's base (Honig and Dinur 1979).

The action of light on rhodopsin

In 1958, Hubbard and Kropf, members of George Wald's team at Harvard, deduced that the only direct effect light has on a visual pigment is that it converts the 11-cis isomer of the chromophore into the all-trans form (Hubbard and Kropf 1958). Dark reactions then follow with the formation of four spectrally well-defined intermediates of increasing lifetimes, the final stage being the splitting of the Schiff's base linkage to give opsin and all-trans retinal. Cis–trans isomerism was universally accepted as the initial stage up to 1972 when irradiation of bovine retinal at 4 K with pulses of only 10^{-12} s (picosecond flash photolysis) showed that the formation lifetime of the first intermediate (bathorhodopsin) was less than 6 ps, a time considered by some to be too short for isomerisation of a large molecule attached to a protein (Busch *et al* 1972) although others reached the opposite conclusion (Green *et al* 1977).

Figure 5.1 The rhodopsin photobleaching and regeneration cycle.

Rhodopsin	(500)	
$h\nu$ ↓ ps		
Bathorhodopsin	(543)	
↓ ns		Enzymes (isomerases)
Lumirhodopsin	(497)	
↓ μs		
Metarhodopsin I	(481)	
↓ ms		
Metarhodopsin II	(380)	
↓ min		
Opsin + all-trans retinal	(381)	

The structures of the four intermediates have not yet been determined and the generally accepted view that all-trans retinal is converted back to the 11-cis isomer by the action of enzymes has still to be proved (Sack and Seltzer 1978). The complete cycle is shown in figure 5.1 the numbers referring to λ_{max} of bovine extracts in digitonin.

It should be mentioned that rhodopsin will be regenerated if *any* of the intermediate products, including all-trans retinal + opsin, are irradiated. At the same time some 9-cis pigment (isorhodopsin) is also formed. This method of regeneration has been omitted from figure 5.1 as it does not occur in the eyes of living organisms under normal viewing conditions. The one exception to this is the UV-sensitive pigment of *Ascalaphus macaronius* which on irradiation forms a MRI of $\lambda_{max} = 480$ nm: this does not regenerate in the dark and the only known method of regeneration is irradiation by visible light (Gogala *et al* 1970).

The structure of rhodopsin

The protein part of rhodopsin has a molecular weight of about 41 000 and the amino acid sequence of bovine rhodopsin has recently been determined and a structural model proposed (Hargrave *et al* 1984a, b). Analysis showed that the molecule consisted of a chain of 348 amino acids with seven stretches of 19–28 units which are hydrophobic in character separated by polar, hydrophilic acids. This suggested that the chain was threaded seven times through the phospholipid bilayer forming the discs in the receptor. For favourable hydrophobic interactions between a protein and adjacent lipid molecules, thermodynamics, non-resonant Raman spectroscopy and circular dichroism measurements suggest that these hydrophobic regions must have an α-helical structure. Two short carbohydrate chains are attached to the polypeptide near the amino end. The lysine side chain which combines with the retinal is number 296 from this end, lying near the centre of the last helical section.

Retinal photoreceptors

In order to be able to begin to understand how the absorption of a photon results in a signal being transmitted along the optic nerve to the brain some knowledge of the basic facts of the neural and electrophysiological aspects of vision is essential. Although the rod cells are not involved in colour vision their structure is simpler and better understood. As the mechanisms involved are believed to be similar to those of the cones only those of the rod, whose essential features are shown in figure 5.2, will be described.

The structure of the membrane surrounding all cells is a double layer of lipid molecules whose hydrophobic ends point towards the other layer and whose hydrophilic ends point either towards the interior of the cell or towards the external fluid. The lipid bilayer is impermeable to all ions but embedded in the membrane are proteins having specific functions, termed intrinsic proteins. All cells contain one such protein which harnesses the energy stored in the phosphate bond of adenosine triphosphate and supplied by the mitochondria to transport sodium and potassium ions across the membrane. They are termed *sodium*

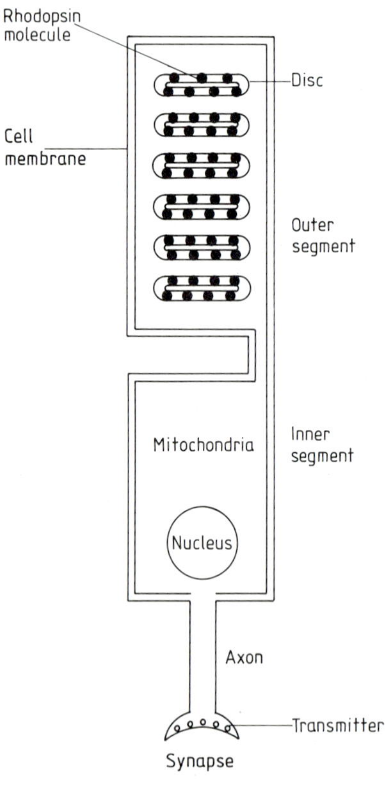

Figure 5.2 The essential features of a rod (diagrammatic).

pumps and their action results in the potassium ion concentration inside the cell being much higher than that outside, the converse occurring with sodium ions. If the membrane itself was permeable only to K^+ potassium ions would flow out of the cell until there was no concentration gradient. As these ions cannot be accompanied by an equal number of anions the inside of the cell becomes negatively charged. This is termed the membrane potential: its magnitude depends on the difference in concentration at equilibrium between K^+ inside (C_i) and outside (C_0) the cell and is given by the Nernst equation which, at 20 °C, simplifies to

$$E_{K^+} = 58 \lg C_0/C_i.$$

The maximum concentration difference gives an equilibrium potential of −75 mV.

If the membrane were permeable only to Na^+ the equilibrium potential would be +55 mV. The membrane surrounding photoreceptor cells is permeable to both K^+ and Na^+ because it contains other intrinsic proteins whose molecular weight is probably in the range 250 000 to 300 000 which serve as K^+ or Na^+ channels. The photoreceptor membrane potential therefore lies between +55 and −75 mV and, in the dark, the rod membrane potential is about −30 mV. If permeability to either K^+ or Na^+ decreases the potential changes towards the equilibrium potential for the other ion. A decrease in Na^+ permeability would therefore increase the potential from −30 mV towards the maximum of −75 mV. Such an increase in potential is termed *hyperpolarisation*: the converse is termed *depolarisation*.

Light absorbed by rhodopsin always causes an increase in the negative membrane potential from the dark value of −30 mV. If only one of the 3×10^9 molecules of rhodopsin in a fully dark-adapted rod absorbs a photon an increase in membrane potential can be detected: it amounts to about 2.5 millionths of a volt (Cervetto and Fuortes 1978) indicating that an extremely powerful method of amplification must occur. Increasing the stimulus causes further hyperpolarisation up to a maximum of −60 mV which occurs when about 100 photons have been absorbed. If the photoreceptor is stimulated between these extremes it will respond to both an increase and a decrease in stimulation unlike conventional nerve cells which exhibit an all-or-nothing response. They do not respond until the stimulus reaches a well-defined threshold when they depolarise and the size of the resultant nerve impulse is quite unrelated to the strength of the stimulus. This conventional all-or-nothing response does occur, however, later in the pathways which lead from the retina to the cerebral cortex.

The mechanism causing hyperpolarisation

It is generally accepted that hyperpolarisation of vertebrate photoreceptors is caused by closing of the Na^+ channels but how this arises from the absorption of a photon by rhodopsin has not yet been fully worked out. In 1972 Hagins suggested that photon absorption induces the release of Ca^{2+} ions from the insides of the discs which then diffuse to the membrane and close the sodium channels (Hagins 1972). Estimates of the number of calcium ions which are released by the disc, however, are far lower than the number required for channel closure (Tyminski *et al* 1982) so there is some doubt about the validity of this hypothesis.

A more promising approach involves a cyclic nucleotide, one of the ubiquitous cellular substances which perform an important role in regulating enzymes. Their possible involvement in photoreceptors was first proposed in 1971 (Bitensky *et al* 1971). The one involved is guanosine 3′5′-cyclic monophosphate (cGMP) and it is believed to be the agent which keeps the sodium channels open in the dark. When rhodopsin absorbs light it causes the release of an enzyme, phosphodiesterase (PDE), molecular weight 240 000. This diffuses to the cell membrane where it hydrolyses cGMP to guanosine 5′-monophosphate (GMP) which is incapable of keeping the Na^+ channels open. How photon-excited rhodopsin releases PDE is not yet known but two other intrinsic proteins are believed to be involved, a nucleotide-binding protein, molecular weight 40 000 (Fung *et al* 1981) and a 'helper' protein, molecular weight 60 000 (Shinozawa *et al* 1980). The rhodopsin intermediate involved is believed to be metarhodopsin II (Fukada *et al* 1981).

The transmission of neural signals

When a conventional receptor is stimulated above its threshold an electrical impulse passes along its axon to its synapse with the next nerve cell, or neuron, in the chain, a mechanism called 'firing'. At the synapse the impulse causes the release of a chemical compound called a *transmitter* which crosses a narrow gap and enters the next cell. The commonest transmitter used by sensory nerves entering the central nervous system is L-glutamate (XXXVIII).

$$\begin{array}{c} CH_2CH_2COOH \\ | \\ H\text{—}C\text{—}COOH \\ | \\ NH_2 \end{array} \qquad \begin{array}{c} CH_2CH_2COOH \\ | \\ H\text{—}C\text{—}H \\ | \\ NH_2 \end{array}$$

(XXXVIII) (XXXIX)

This acts on a specific intrinsic protein in the cell membrane which controls the opening of ion channels which permit the entry of sodium and other positively charged monovalent ions. This influx causes depolarisation which generates a positively charged impulse to the next synapse where this in turn releases a transmitter which then enters the next cell. This process is repeated until the signal reaches its terminal in the brain. L-glutamate is therefore termed an 'excitatory transmitter'. Most non-receptor neurons, however, also receive an 'inhibitory transmitter' which has the opposite effect. The commonest in the central nervous system is γ-aminobutyric acid (XXXIX) known as GABA. This acts on another intrinsic protein which controls other ionic channels, those permitting negative chloride ions to enter. This causes hyperpolarisation and has the effect of inhibiting any firing signal sent by another neuron. Both of these transmitters, together with many others, have been found in the retinas of many species. Some of these have been found to be produced by horizontal and amacrine cells but the neurotransmitter(s) produced by rods and cones have not yet been identified (Shichi 1983).

Signal transmission and processing in the primate retina

Non-vertebrate photoreceptors behave like conventional receptors: when the light falling on them exceeds threshold intensity they fire and an excitatory transmitter is released at the synapse. We have seen, however, that rods and cones do not fire but a change in the intensity of the stimulus causes a change in the membrane potential. This affects the *rate* of transmitter release at its synapse with the next neuron, a bipolar cell. Each cone is synapsed to at least one bipolar and in the centre of the fovea, where colour vision is most discriminating, the number is *three* (Polyak 1957, Vilter 1949). Each bipolar receives signals from more than one cone of the same type, the number increasing away from the fovea resulting in lower acuity. The rate of transmitter release by the cone governs the potential of the bipolar but two separate mechanisms are involved.

The simplest mechanism results in the bipolar potential changing in the same way as that of the cone, i.e. an increase in stimulus intensity causes hyperpolarisation of both the cone and the bipolar. About half of the bipolars, however, respond in the opposite way, i.e. stimulus increase causes depolarisation (Werbling and Dowling 1969, Kaneko 1970). It is easy to see the advantage of this: sending opposite signals along separate channels will increase the range of effective stimulus intensities.

Neurons of a given type in receptor pathways only produce one transmitter, excitatory or inhibitory, and it is therefore difficult to see how the release of one transmitter can have opposite effects. Until the actual mechanism is known the effect is conventionally attributed to the nature of the synapse which is called 'sign conserving' if the post-synaptic cell responds in the same way as the pre-synaptic cell (in this case a cone) and 'sign inverting' if the opposite effect occurs.

The bipolars are synapsed to ganglion cells whose axons constitute the optic nerve leaving each eye. Ganglion cells do not exhibit graded potentials but fire in the conventional manner sending signals to the cerebral cortex. The receptor-bipolar-ganglion pathway is often termed the *vertical organisation* of the retina and there is a very much more complicated *horizontal organisation*. This involves horizontal cells which carry signals laterally at the level of the receptor-bipolar synapses and amacrine cells which perform the same function at the bipolar-ganglion synapses.

Before we consider the complicated responses of ganglion cells, however, it is interesting to consider the first recordings made from neurons exhibiting graded potentials which were made by Svaetichin in 1953 on cells in the retina of the carp.

Fishes have been of immense value to neurophysiologists because they possess colour vision and their retinal cells are much larger than those of primates. Svaetichin exposed the carp to flashes of monochromatic light, varying the wavelength, and found that the cells fell into two major classes. One gave a hyperpolarising response having a broad spectral maximum which he termed the luminosity response. The other gave a hyperpolarising response over part of the spectrum and a depolarising response over the remainder which he termed a chromatic response. He immediately recognised that this provided a physiological basis for the opponent-colours theory and he dedicated his paper to Hering who had not lived to see his theory vindicated (Svaetichin 1956). The discovery of a single cell giving opposite responses to different stimuli had revolutionary implications: neurophysiologists had rejected the opponent-colours theory because it was an article of faith with them that a neuron was either quiescent or else it fired if it was stimulated above threshold (Grundfest 1958).

It is interesting to note that Svaetichin originally believed that his recordings had actually been made from the cones themselves and had this been true it would have invalidated the trichromatic theory, at least for the carp. MacNichol and Svaetichin, however, repeated the experiments using a micropipette electrode containing a dye which diffused into the cell but not through the membrane: the actual cell from which the recordings had been taken was therefore stained and could be much more reliably identified by subsequent microscopic examination of a cross section. They found that the recordings had been made on horizontal cells (MacNichol and Svaetichin 1958) which must have received signals from both long- and short-wave cones, one via a sign-conserving synapse, the other via a sign-inverting synapse.

The responses of ganglion cells

When a microelectrode is placed against a ganglion cell in the dark, pulses occurring at a low, fairly steady rate, *the resting rate*, can be recorded as spikes on a CRT and as clicks in a loudspeaker. When a small spot of light is shone on to the retina a well-defined position can be identified where the *change* in firing rate is at a maximum. This position is the centre of the *receptive field* (RF) of that cell and it may be the position of maximum or minimum rate of firing. Cells of the former type are usually termed 'on-centre' and designated '+' those of the latter 'off-centre', designated '—'. In primate retinas these cells are of two types, *tonic* and *phasic* but as the latter only respond momentarily at the beginning or end of a sustained simulus they need not be considered as their involvement in *colour* perception is most unlikely.

The simplest type of primate ganglion cell responds in the same sense to light of all wavelengths falling on the centre of the RF with a broad spectral maximum. As the spot moves radially outwards the response changes to that of the resting state but then changes in the opposite sense, reaches a maximum or minimum and then returns again to the resting state. This centre/surround effect is termed 'spatial opponency' and these spectrally non-opponent cells are probably involved in the perception of lightness.

Spatial opponency is a characteristic of almost all (>96%) ganglion cells and most (>80%) are spectrally opponent as well, the most common exhibiting an increase in firing rate when red light is shone on to the centre of the RF and a decrease when green light is shone on to the surround. Such a cell is described as 'red on-centre/green off-surround' and will be abbreviated to $R+/G-$. There are five types of spatially and spectrally opponent cells in primate retinas (Gouras and Zrenner 1981) namely

$$R+/G-$$
$$G+/R-$$
$$R-/G+$$
$$G-/R+$$
$$B+/Y-.$$

The complete absence of $B-/Y+$ cells implies that the blue cone channel only utilises on-centre mechanisms but why this should be the case is a complete mystery.

Colour vision mechanisms in the brain

The axons of the ganglion cells form the million or so fibres of the optic nerve leaving each eye. These appear to cross at the *optic chiasma* and then proceed to the left and right hemispheres which are so characteristic of the brain. Isaac Newton, however, correctly deduced that only about half of the fibres actually cross, those coming from the nasal half of each retina. The two bundles of fibres beyond the chiasma, known as *optic tracts*, pass upwards through the white matter of the brain and enter the left and right *lateral geniculate nuclei* (LGN). Each LGN therefore receives signals from both eyes and the cells which are synapsed to the fibres of the optic tracts are arranged in alternate layers, three for each eye. The responses of these cells appear to be identical with those of the ganglion cells and it is tempting to regard the LGN as merely relay stations on the way to the next stage, the cortex. This is, however, unlikely to be the case because retina–cortex distances are far too short: there are many nerve fibres over two feet in length. The true function of the LGN has yet to be discovered.

The cortex, the grey matter of the brain, consists of a deeply folded plate of neural tissue about 2800 cm^2 in area and about 2 mm thick. It is split down the middle, each half completely covering one of the brain's hemispheres. By the beginning of the century it was recognised that different parts of the cortex had different functions and that those involved in vision were the well-defined areas situated at the back of each hemisphere. Each of these areas receives signals from its own LGN and the partial crossing of the optic nerves causes that on the left hemisphere to receive information about the centre and right-hand side of the visual field from both eyes, with information from the centre and left-hand visual field going to that on the right hemisphere. Both areas are therefore always treated as one, known variously as the *primary visual cortex*, *striate cortex* (because of the striped appearance of cross sections under the microscope), *Area 17* (as designated on Brodman's 1905 map of the brain) and *Area V1* (Zeki 1971).

The concentration of neurons in striate cortex (the definite article is usually omitted) is extremely high – about 100 000 per square millimetre of surface – and in 1959 Hubel and Wiesel began making recordings from individual neurons first in the cortex of the cat and later (1968) in that of the rhesus monkey. This resulted in an astonishingly detailed understanding of the mechanisms involved and for this work they were awarded the Nobel Prize for Medicine in 1981. Their account of this work is extremely readable (Hubel and Wiesel 1979).

These cortical neurons are arranged in eight layers and most of the cells which are synapsed to the axons of LGN cells lie in layer 4C. Like the LGN and ganglion cells they have spatially opponent concentric centre/surround RFs and form a map of the retina varying in scale as it does in a Mercator projection; the scale for the fovea (where visual acuity and colour discrimination are maximal) is about 35 times larger than that for the peripheral retina. Cells above and below layer 4C all respond best to linear stimuli of specific orientation, their RFs varying in complexity and termed simple, complex and hypercomplex cells. One prominent feature of the distribution pattern of these cells is that they are arranged in columns, all the cells in a column having one visual feature in common. The features most clearly showing columnar organisation are orientation (though not, of course, concentric cells) and ocularity dominance, i.e. wholly or mainly driven by the same eye.

In view of the fact that the majority of LGN cells are spectrally opponent, Hubel and Wiesel expressed surprise at finding only a few cells in striate cortex which were wavelength selective (Hubel and Wiesel 1968). Some of these had spectral responses identical with LGN cells but the most interesting exhibited two opposite kinds of spectral opponency. Such cells are termed *double opponent* and many were later found by Michael (1978). One type gives an 'on' response to red light and an 'off' response to green light falling on the centre of the RF and the opposite responses when light falls on the surround, i.e.

$$R+G-/R-G+.$$

$G+R-/G-R+$ cells were also found but so far no double-opponent cells receiving an input from blue cones have been found. This may be because no $B-/Y+$ cells have been found in primate retinas.

Simple cells exhibiting double opponency were also found but complex and hypercomplex cells only exhibited an 'on' response, either $R+$ or $G+$, the spectral sensitivity being similar to that of the 'on' response of the double-opponent cells in the same column. These cells, unlike the double-opponent cells, were also driven by both eyes.

There seems little doubt that the synaptic organisation within a column is serial – concentric to simple to complex to hypercomplex – and that these cells then send signals (project) to other parts of the brain. The complex cells in layer 6 project mainly (perhaps only) back to the LGN; those in layer 5 project to the *superior collicus*, a visual station in the white matter (Hubel and Wiesel 1979).

The part most extensively studied, however, is the adjacent area of the cortex which, because it lies in front of striate cortex, is usually known as the *prestriate cortex*. It includes Areas 18 and 19 on Brodman's map and has long been assumed to be involved in the interpretation of the visual scene which has been analysed in striate cortex. It receives signals from most layers of striate cortex and its cells are characterised by having much larger RFs than those in striate cortex. Its surface is not another map of the retina but consists of three areas having different functions which Zeki designated V2, V3 and V4 with V1 being his designation of striate cortex (Zeki 1971). Conventional techniques for tracing connections between neurons – one uses HRP, the delightfully named *horse-radish peroxidase* – have shown that cells in V2 receive their input from the double-opponent cells in layers 2 and 3 of striate cortex (Livingstone and Hubel 1984) but so far no recordings have been made from cells in V2 so their function is unknown. They, in turn, project exclusively to Area V4 (Shipp and Zeki 1984) where Zeki (1973) had made recordings from 77 cells in eight monkeys and found that each responded 'vigorously to one wavelength and grudgingly, or not at all, to other wavelengths or to white light', there being approximately equal numbers of on and off responses (Zeki 1973).

Further recordings made from 62 cells showed that each was responsive to only a narrow spectral region, many having a bandwidth at half maximum of between 10 and 20 nm. These covered the whole spectrum and the extra-spectral purple region with maximum frequencies at 480, 500 and 620 nm. No cells were found which were responsive to yellow light at 560–570 nm, the peak of *R* cone sensitivity (Zeki 1980). It would seem very likely that this is connected with the fact that this is the spectral region of minimum perceived saturation or chroma. He also found that when the field consisted of a multicoloured display the response of some of the cells was not governed by the composition of the light reflected from the area in the centre of the RF but by its colour as perceived by the investigator. The significance of this finding will become apparent when the retinex theory of colour vision is described in chapter 11.

Other investigators, however, have failed to confirm that Area V4 is particularly rich in colour-coded cells (Krüger and Gouras 1980, Schein *et al* 1982). Krüger and Fischer (1983) have suggested that the fact that Zeki's monkeys were both anaesthetised and paralysed may have suppressed signals from non-retinal sources which normally govern the responses of these cells.

Another complicating factor is that the long-held belief that prestriate cortex received signals only from striate cortex is wrong. In 1967 it was discovered that the cat prestriate cortex received signals directly from the LGN (Wilson and Cragg 1967) and this was later found to be the case in New World monkeys (Wong-Riley 1976) and the macaque monkey (Yukie *et al* 1979, Fries 1981).

Models exhibiting colour vision mechanisms

Several such models have been devised of which two are worthy of special mention. Gouras and Zrenner (1981) have published a schematic circuit of on- and off-centre pathways involving cones, bipolar, horizontal, amacrine and ganglion cells. Hunt (1982) has devised a circuit diagram involving linear and non-linear summations and differentiations of cone outputs which provide good predictions for loci of constant hue, simple criteria for unique hues and good approximations to the Natural Colour System and Munsell schemes for surface colours and to data on the appearance of spectral colours.

There is no doubt, however, that these and other cone-to-cortex circuit diagrams will need major revisions as further studies reveal new facets of the remarkably complex mechanisms involved in the perception of colour.

The rivalry between the two main theories ended, of course, as soon as spectrally opponent cells had been discovered in the primate LGN. The trichromatic theory explained the first stage of the visual process, the opponent-colours the second. Such two-stage theories had actually been put forward during the controversy (Kries 1905, Adams 1923, Schrödinger 1925, Müller 1930) and two such theories were used to derive colour difference equations, the subject of chapter 10.

6

Defective Colour Vision

Approximately eight per cent of the male population have impaired colour vision which is commonly, but misleadingly, called colour blindness. The incidence among females is only about one half of one per cent, the cause of this marked difference lying in the genetic mechanism involved, the gene responsible for colour blindness being located on the X sex chromosome and is thus dominant in the male and recessive in the female. The first detailed account of inherited colour vision defects was given by the chemist John Dalton in 1798 who, together with his brother, could not differentiate colours in the red-yellow-green region of the spectrum and to whom bright red colours appeared dark. Thomas Young (1807) suggested that the cause of this defect was the lack of the red cone pigment and the defect was named *protanopia.* Another class of red-green confusers did not perceive red colours as being dark and this defect was assumed to be caused by the lack of the green cone pigment: this defect was termed *deuteranopia.* A third class appeared to lack the blue cone pigment and this class was naturally named *tritanopia.* In 1952 the widely circulated magazine *Picture Post* published a confusion chart (see below) designed to detect tritanopia (Wright 1952), this paper being a valuable addition to his earlier publication (Wright 1946). By applying his technique of retinal densitometry to protanopes and deuteranopes, Rushton obtained convincing evidence that this explanation based on the trichromatic theory, was correct (Rushton 1975). The absence of one or other of the three cone pigments means that instead of requiring three additive primaries for colour matching these observers only require two and are collectively termed *dichromats.* As the visual acuity of protanopes and deuteranopes is not impaired it is assumed that they have a full complement of active cones but that there is only one pigment in addition to the B cone pigment. The protanope has only the pigment present in the G cones of the normal individual and in consequence the far red of the spectrum appears much darker. The deuteranope has only that pigment in the R cones of the normal individual. This does not cause any reduction in the relative brightness of the green region of the spectrum because of the marked overlap of the R and G cone sensitivity functions (figure 4.1).

Confirmation of the single long-wavelength cone hypothesis has also been obtained by microspectrophotometric measurements on the excised retina of a

75 year old male deuteranope (Mollon *et al* 1984a). These were made on 17 receptors, none of which had an absorption maximum in the region 527.3–534.3 nm which is characteristic of green cones (Dartnall *et al* 1983). Mollon *et al* comment that this was not only the first measurement made on the cones of a dichromat, it was also likely to be the last because ophthalmologists are becoming reluctant to remove eyes which have developed small melanomas.

Protanopia and deuteranopia each occur in about one per cent of the male population; tritanopia is much rarer – 0.002% male and 0.001% female (Wright 1952). Because dichromats experience severe difficulty in naming colours correctly, they are unlikely to apply for positions involving colour perception.

At the turn of the century Lord Rayleigh discovered that most observers required approximately the same proportion of red and green lights to match a given yellow but that some required significantly different proportions. These observers were trichromats and they were called *protanomalous* if they required more red than normal and *deuteranomalous* if the opposite were true. The incidence of protanomaly amongst the male population is 1.0%, that of deuteranomaly 4.9%. Though their colour vision characteristics are obviously different from normal this does not, of necessity, mean that their colour vision is defective but in practice this always turns out to be the case as they cannot repeat their matches with the same degree of precision as a normal observer and their ability to perceive colour differences between colours of low saturation is impaired. The impairment varies from slight to a severity approaching that of dichromacy. Anomalous trichromacy is believed to be caused by the *R* cone pigment having a blue-shifted λ_{max} (protanomalous) or by the *G* cone pigment having a red-shifted λ_{max} (deuteranomalous). In either case the absorption curves will have greater than normal overlap and hence the discrimination will be impaired.

Extensive behavioural experiments have shown that colour vision defects similar to those occurring in humans also occur in monkeys. All the limited number of studies on Old World monkeys have not shown any marked divergence from trichromacy but one New World monkey, the squirrel monkey (*Saimiri sciureus*) exhibits the most common variants of human dichromatic and trichromatic colour vision: females are either dichromats or trichromats whereas all males are dichromats (Jacobs 1984).

Ten such monkeys were recently sent from California to Queen Mary College, London, and microspectrophotometric measurements made on their photoreceptors after death. All retinas contained rod cells (λ_{max} approximately 500 nm); some blue cones were found and in the medium- to long-wavelength part of the spectrum there were at least three photopigments with λ_{max} values about 537, 550 and 565 nm, each animal having either one (dichromat) or two (trichromat). In a double-blind experiment the type of colour vision predicted from microspectrophotometric measurements made in London gave excellent agreement with that indicated by the behavioural studies made in California (Mollon *et al* 1984b).

The colour sensations of dichromats

To an observer with normal colour vision every part of the spectrum appears to be coloured. Every dichromat, however, perceives a narrow region as colourless (achromatic) and the wavelength of the neutral point can be used to identify the type of dichromat as follows:

Protanopes	490–495 nm	(Walls and Heath 1956)
Deuteranopes	495–505 nm	
Tritanopes	568–570 nm	(Cole *et al* 1966).

As the wavelength departs from the neutral point one colour sensation is experienced at shorter wavelengths and an entirely different colour sensation at longer wavelengths. One fascinating problem has been the identification of the sensations experienced by protanopes and deuteranopes. Early advocates of the trichromatic theory had no doubts at all: the hue sensations of protanopes are blue and green, those of deuteranopes, blue and red. Supporters of the opponent-colours theory, however, explained protanopia and deuteranopia as being caused by the loss of the R/G process which would mean that both types would perceive the long wavelength part of the spectrum as yellow. It is, of course, pointless to ask them what colour sensation they experience: by varying the brightness of long wavelength monochromatic light against a constant white background responses of red, yellow or green will be obtained. It might be thought that this question (of colour sensation experienced) was unanswerable but a possibility lies in the existence of unilateral dichromats, observers possessing one dichromatic and one normal eye who can make binocular matchings. The wavelength selected using the normal eye should give a valid answer. Most of the evidence from unilateral dichromats seems to indicate that their long wavelength sensation is yellow but there is always some doubt that the dichromatic eye exhibited typical protanopia or deuteranopia so an unambiguous answer has not yet appeared.

The identification of colour-defective observers

In the colorant manufacturing and using industries personnel involved in visual colour matching must have normal colour vision and in view of the incidence of defective colour vision among males (about one in twelve) some screening test is essential (McLaren 1966). Fortunately all this test has to do is to divide the population into normal and defective. It is not necessary to know the nature of the defect nor its severity as even an anomalous trichromat whose defect is only mild will be unable to differentiate between a reddish grey and a greenish grey. The method which has withstood the test of time (since 1917) is the set of confusion charts devised by Ishihara (1973). Plate No 2 in this reference, for example, consists of a figure 8 outlined in bluish red or brown spots against an olive-green spotted background which is clearly perceived as an 8 by any observer with normal colour vision. The brown spots appear on the left-hand side of the figure and these cannot be differentiated from the olive-green background by protan or deutan anomalous trichromats or dichromats who unhesitatingly state that the number seen is 3. No tritan confusion plate is included.

Three precautions are essential. Firstly, the test must be carried out under north sky daylight in the northern hemisphere or preferably under a fluorescent lamp conforming to BS 950: Part 1. Secondly, the number of errors an observer was permitted to make before being classed as defective was not specified in editions 1-14 and in the past many testers would play safe by rejection at the one or two error level. This was remedied in the 15th (1960) and subsequent editions and any observer making fewer than five errors on Plates 1-21 must be classed as possessing normal colour vision. Thirdly, the tester must always remember that even if an observer makes seventeen or more errors he (very rarely, she) may only be an anomalous trichromat who has never had any difficulty in correctly naming any colour nor has he ever made a bad mismatch. He should not, therefore, be told that he is colour blind but that his colour vision is not up to the highest standards demanded. During the Second World War there were several instances of observers having been rightly rejected for industrial colour matching against the colours of low saturation in the Ishihara plates who have subsequently passed the lantern tests used to test pilots.

Because of its severity, the Ishihara test is not a suitable screening test for those occupations requiring good, but not excellent colour discriminating ability, for example, electronic engineering where colour-coded components are used. In such cases a test capable of grading the severity of the defect must be used and the excellent monograph by Voke (1980) or the more comprehensive book by Fletcher and Voke (1985) should be consulted.

Yellowness of vision

Although not in any way a colour vision defect, variations in yellowness of vision can be a major cause of disagreements between colour matchers and therefore their degrees of yellowness should be determined as a matter of routine. The fovea, the central region of the retina involved in critical visual tasks, is permeated by a yellow carotenoid pigment, xanthophyll, and the intensity of pigmentation varies between individuals from negligible to an extent which reduces the amount of light at 460 nm transmitted to the cones by two-thirds. A second cause of variations in yellowness of vision is yellowing of the lens itself and, unlike the extent of macular pigmentation, this increases with increasing age. Over the normal working life span (16 to 65) differences caused by lens yellowing are about twice as great as the variations within any one age group which can be assumed to be due to differences in macular pigmentation.

Though one would expect observers with maximum yellowness of vision to be less sensitive to colour differences caused by differences in reflectance at the blue-violet end of the spectrum, there is no evidence that this is the case providing that the level of illumination is sufficient to compensate for the increased absorption. Significant differences between observers whose yellowness of vision varies only appear when a certain type of colour match is being assessed. The first stage of the colour vision process is trichromatic and two stimuli which cause the same degree of excitation of each of the three cone types will be perceived as identical. If the spectral compositions of the two stimuli are the same the match may be termed a 'spectral match'. If they are different, the match is called 'metameric' (Ostwald 1919). Spectral matches are therefore usually called 'non-metameric'. The mechanism by which two spectrally different stimuli can cause the same degree of excitation of each of the cone types can be seen by reference to figure 6.1. The curve is an idealised absorption curve for one cone type and the two straight lines are idealised spectral power distributions (SPD) of the light entering the cone from the standard and from the batch. It will be seen that over half of the spectral range absorbed by the cone, the SPD of the standard is greater than that of the batch but the situation is exactly reversed over the other half. The effect on the cone is given by the summation of the SPD at each wavelength multiplied by the height of the absorption curve at the same wavelength and it will be readily seen that in this case the two different SPD will give identical summations. Symmetry was introduced solely for ease of understanding: identical summations will be given by a very wide (theoretically infinite) range of SPD the only essential requirement being

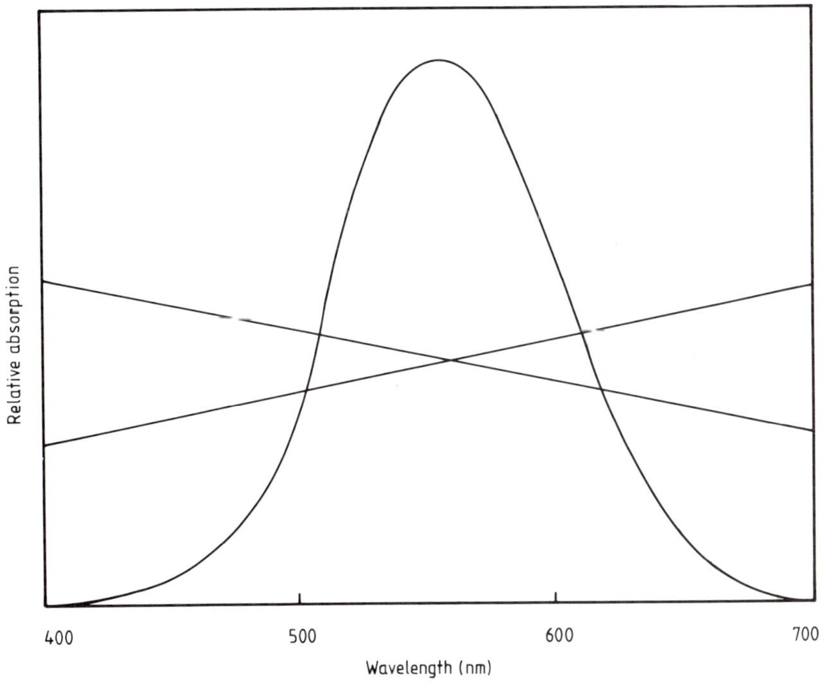

Figure 6.1 The origin of metamerism (see text).

that the curves cross at at least one point. Identical summations for the other two cone types as well will be given by an equally wide range of SPD, the only essential requirement being a minimum of three crossover points: an analysis of 124 metameric pairs showed that the average number of crossover points was 4.4 (Berger and Strocka 1973). As the SPD causing identical excitations of each cone type are in part governed by the absorption by the lens and the macular pigmentation it is easy to see that identical excitations will not occur if there are differences in yellowness of vision and hence the two specimens will not always be a match.

The increase in yellowness of vision with age is well known to dyers and many young dyers whose matches are going to be subsequently inspected by an older colourist will attempt to allow for this by producing a batch which, to them, requires the addition of a small amount of yellow dye to bring it on shade. This, of course, will make matters worse if the match is non-metameric because all observers will see that yellow shading is required. Unfortunately, even if the match is metameric, this ruse will make matters worse as often as it will succeed because the colourist cannot predict, even qualitatively, what differences he will see between standard and batch which the older observer sees as a perfect match. The reasons for this are as follows. If two specimens are a match to the older observer his B cone responses must be identical. If the reflectance curves of these specimens are not identical over the spectral region of B cone sensitivity then the B cone responses of a younger observer with a lower degree of pigmentation cannot be the same: one specimen will therefore appear bluer than the other, i.e. will require shading with a yellow. If this specimen is the batch, the ruse will succeed but it is just as likely to be the standard when it will fail.

Fortunately it is easy to assess the degree of yellowness of vision and if every colourist knows his or her degree of yellowness then those differing markedly will realise that arguments are a waste of time. There is no point in trying to define the correct degree of yellowness. In practice, the correct degree is always that of the observer having the power to accept or reject the batch as a good commercial match.

The determination of yellowness of vision

The first method for determining the degree of yellowness of vision was developed by James Glenn of Sidney Blumenthal and Co. Inc., Shelton, Connecticut, USA in 1941 and improved by F T Simon in 1943. This method consisted of two 24-step scales of dyed wools, one going from blue to brown, the other from violet to yellow-green†.

In 1967, Davidson and Hemmendinger produced a similar device using painted cards and this version is the one which has been most widely used‡. The first scale consists of 21 painted areas varying in steps from violet (A), through

† Obtainable from AATCC, Research Triangle Park, North Carolina, USA.

‡ Obtainable from Munsell Color Co., PO Drawer 950, Newburgh, NY, 12550, USA; Kollmorgen (UK) Ltd., Bridgewater House, Sale M33 1EQ.

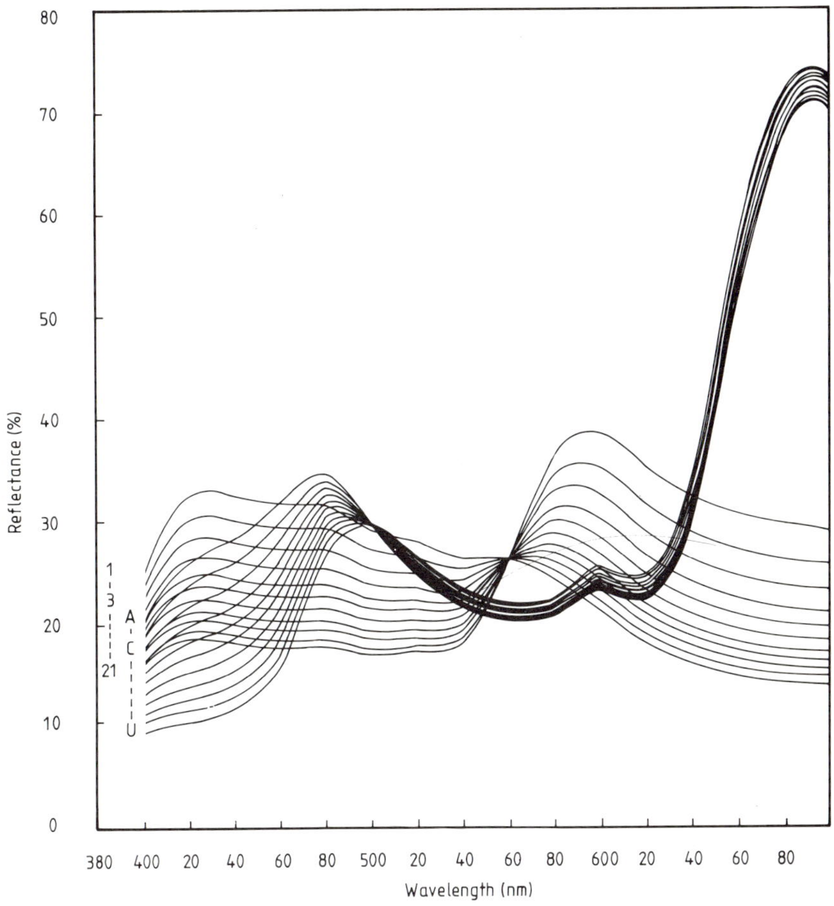

Figure 6.2 Reflectance curve of alternate areas of the Davidson and Hemmendinger Color Rule.

grey to olive green (U); the second scale varying from blue (1), through grey to brown (21). Reflectance curves of alternate areas are shown in figure 6.2. To assess the yellowness of vision the observer views the two scales under a standard daylight lamp and selects the pair, one from each scale, which gives the closest match. Fifty-nine observers whose ages ranged from 7 to 64 years gave matches ranging from E7 for the observer possessing least yellowness to L13 for the one with the greatest degree of yellowness (Kaiser and Hemmendinger 1980).

Once two colourists have seen that each can repeat his/her matches but that the choice of each is a mismatch to the other, they realise the futility of any arguments over metameric matches.

7

The Variables of Perceived Colour Vision: Colour Atlases

Colour is a word that does not really require a definition and, in fact, the only meaningful definition for everyday speech is that it is what we mean when we use words such as red, purple, brown etc but not black, grey or white. In colour science, however, these last three sensations are regarded as colour sensations in their own right and they are differentiated from the former by saying that they do not possess the attribute of *hue*. They are therefore achromatic sensations. Hue is therefore the most important variable of perceived colour and it is also the least complex as hue is the only term ever used in colour science to describe this attribute. It is also a word used in everyday speech with exactly the same meaning.

There are two other independent attributes of perceived colour each of which is defined differently by different terms: in this chapter we shall follow the definitions in the CIE† vocabulary as far as possible but some are, unfortunately, being re-defined or introduced for the first time.

The first of the two remaining variables is that of *lightness* which, in the case of object colours, is that attribute according to which an area appears to reflect diffusely or transmit a greater or smaller fraction of incident light. The last variable is the most complex of all and this is well illustrated by the proliferation of terms used to describe it. The one most familiar to dyers and pigment technologists is probably *saturation*, other terms being *intensity*, *purity* and *vividness*.

The fact that there are three independent variables of perceived colour explains why it is impossible to arrange a random selection of colours in a logical order in a dye or paint manufacturer's shade card. Such a shade card is only a two-dimensional array even if arranged in book form. The most logical way of ordering colours was actually discovered by the Swedish monk and astrologer Sigrid Aron Forsius in 1611, before Newton was born! His array was based on a central vertical axis to represent the variable of lightness with black at the bottom and white at the top. The hues were arranged around this axis in a complete circle. The various colours occupy a three-dimensional colour space

† Commission Internationale de l'Eclairage, 52 Blvd. Malesherbes 75008, Paris, France.

and such an array is known as a *colour order system*. Several have since been devised and they all resemble that of Forsius in having a central, vertical achromatic axis but they differ in the organisation of the other two variables. Their importance in colour science and technology is indicated by the fact that an international symposium on the subject was held in Sweden in 1983 (AIC 1983).

The most important colour order systems are described in the following paragraphs.

Munsell colour space

Munsell illustrated his concept of colour space by painting colour chips which were equally spaced in each of the three dimensions. He termed the lightness variable *Value* and one Value step was visually equal to two *Chroma* steps. This word, coined by Munsell, is now adopted (with a small c) by the CIE. The complete hue circle was divided into five principal hues red, yellow, green, blue and purple with several intermediate hues designated by a mixture of numbers and letters, e.g. 5YR for orange hues. Munsell had a deep-seated aversion to using names of objects, for example orange and violet, as colour names. These colour chips were stuck on charts and published as the first colour atlas in 1915.

The spacing of the chips was intensively studied by the Colorimetry Committee of the Optical Society of America and in 1943 the CIE tristimulus values (see chapter 8) of ideally spaced chips were published as the Munsell Renotation System (Newhall *et al* 1943), which has been indispensable for the development of a method of objectively quantifying the variables of perceived colour. The Munsell colour system has thus successfully bridged the gap between art and science.

The current Munsell atlas, sold as the *Munsell Book of Color*†, is published in two parts, glossy (1488 chips) and matte (1277 chips). For understanding the geometrical attributes of colour space, the Munsell Color Tree illustrating over 200 glossy chips is ideal: it is shown in plate 1. This logical representation of colour space is a three-dimensional Euclidean space expressed in cylindrical co-ordinates. The major defect of such an array concerns the illustrations of hue differences since two chips on adjacent hue plates illustrate a very large hue difference for colours of high chroma but a vanishingly small difference as the chroma approaches zero. This is undesirable as there are too few chips illustrating high chroma colours which are industrially far more important than those of low chroma.

The OSA-UCS Specimens‡

To overcome the difficulty posed by the lack of chips illustrating high chroma colours the Optical Society of America (OSA) spent thirty years producing a

† Obtainable from The Munsell Color Co., PO Drawer 950, Newburgh, NY 12550, USA and The Tintometer Ltd, Waterloo Road, Salisbury, England.

‡ Obtainable from the Optical Society of America, 1816 Jefferson Place, Washington DC.

series of equally spaced specimens. The basis for this arrangement was a cubo-octahedron, a cube whose corners have been cut away by planes passing through the centres of the three edges forming each corner. Each of the twelve new corners is the same distance from each of its nearest neighbours and from the centre of the original cube. If each corner then becomes the centre of another cubo-octahedron, any volume can be packed with no spaces remaining. The OSA produced 558 specimens located at these positions starting from a neutral grey of Munsell Value 6 (Nickerson 1981).

The Natural Colour System†

The term 'natural colour system' was coined by Hering to describe the colour order system based on his opponent-colours concept which was described in chapter 4. This postulated that every surface colour could be described in terms of six elementary attributes – whiteness (w), blackness (s), redness (r), green-ness (g), yellowness (y) and blueness (b). Experiments carried out at the Swedish Colour Centre Foundation showed that observers are capable of making quantitative assessments of these elementary attributes and the basic system was refined and quantified so that for every surface colour: (*a*) the sum of the elementary attributes is always 100; (*b*) the colour can have at the most four elementary attributes and (*c*) the colour can have at the most two chromatic elementary attributes and that these cannot be redness and green-ness nor yellowness and blueness (Hård and Sivik 1981). The system is published as Swedish Standard SS 01 91 00 and is illustrated by a colour atlas containing some 1300 colour samples (SS 01 91 02). CIE co-ordinates for 16 000 natural colour system colour notations are available as SS 01 91 01.

Chroma Cosmos 5000‡

The major weakness of the systems just described is that only a comparatively small number of different colours are illustrated. The Japanese Chroma Cosmos follows the Munsell notation but extends the number of colour chips from about 1500 to 5000 (Birren 1979, Kelly 1981).

The ICI Colour Atlas

Even 5000 specimens, however, are still only a very small fraction of the colours a textile dyehouse might be called upon to produce: the sewing thread firm J and P Coats of Paisley, for example, has no fewer than 10 000 colour standards. To cater for the needs of industry ICI produced a *Colour Atlas* which illustrated over 27 000 different colours. 1379 of the brightest colours obtainable with non-fluorescent light-fast pigments were printed in order on to twelve cards and each of these could then be overlaid by one of twenty neutral grey filters. A

† Obtainable from the Swedish Standards Institution, Stockholm.

‡ Obtainable from the Japan Colour Research Institute, 1–19 Nishiazabu 3-chome, Minato-ku, Tokyo 106.

simple letter/number code identified the combination which any other person with access to an atlas could then set up (McLaren 1971a).

The SCOTDIC System†

Colour atlases which illustrate colours by painted or printed cards have only a limited appeal to textile designers because their texture is so different from any textile. To overcome this defect the Kensaikan Company in Japan produced 2450 shades of dyed polyester and 2000 shades of dyed cotton in various formats (ring book, small swatches, large pieces). Each shade is given a modified Munsell notation. This collection is known as the Standard Colours of Textiles (SCOT) and is accompanied by the international dictionary of colour devised by DIC s.a. of Paris which became part of Kensaikan in 1982. This is a list in alphabetical order of some 2000 colour names (including Japanese names) and their modified Munsell notations.

The ISCC-NBS System

The ISCC-NBS system partitions the whole of Munsell colour space into 267 blocks, the centre of each block being illustrated by a painted chip and given a simple colour name consisting of not more than four words, for example very deep purple, light reddish grey. It is accompanied by a dictionary of some 7500 international colour names arranged in alphabetical order giving the number of the nearest centroid colour. Another index lists all of the names referring to each centroid colour. It is, unfortunately, rarely used probably because it is based on Munsell co-ordinates which are not easily obtained from CIE tristimulus values (McLaren 1983a).

The Visual Color System (VCS)

The basic principle behind the Visual Color System invented by Ralph Stanziola when he was with Applied Color Systems Inc. of New Jersey, is Maxwell's early method of generating various colours by spinning a disc consisting of red, green and blue segments. Seven segments are used in the VCS – black, white, red, yellow, green, blue and violet – which give a much greater colour gamut. The amounts of each colour can be varied whilst the disc is spinning and when the desired shade is obtained these amounts are displayed digitally so that the same colour can be generated on any VCS instrument. When interfaced to an ACS colour computer recipes to obtain the shade are generated.

The multiplicity of colour atlases

The fact that there are so many colour atlases, augmented by collections of colour samples arranged in no logical order such as the 'Pantone' collection widely used in the non-textile printing industry, causes many problems in communication. It has been suggested that these could be largely overcome if every

† Obtainable from Tom Lewis Studios Ltd, 47 Kennerley Road, Stockport SK2 6EU, England.

illustrated colour was identified by its CIELAB lightness, chroma and hue co-ordinates so that cross-referencing would be easy. This would also be applicable to the colours generated by the VCS system and also to those on a VDU once there is agreement on the lightness level of the grey surround (McLaren 1985b).

The Munsell variables of perceived colour

The ability of untrained observers to quantify the lightness, chroma and hue components of a random collection of ten coloured cards was shown by an experiment carried out at an exhibition in 1971 in which 323 observers took part. Their estimations showed a surprising degree of precision except in the case of the specimen which was dark brown. In this case although the standard deviation of the lightness estimates was the lowest of all, those for hue and chroma were over three times larger than the means for the other colours. What was even more significant was the fact that a quarter of the observers found it impossible to analyse the hue in terms of the four psychological primaries while a further 11% reported that it was extremely difficult (Padgham and Rowe 1973).

The uniqueness of brown

After words for the four psychological primary hues have been added to a language, the next one is usually a word for the sensation described in English as *brown* (Berlin and Kay 1969) which indicates that this colour has an importance greater than those described as orange, grey, purple, etc.

Objects coloured with dyes or pigments are normally seen in relation to other colours and the opposite situation occurs when a stimulus is perceived in isolation against a perfectly uniform black background most simply, but not exclusively, produced by viewing a light source in a completely darkened room. There are certain colour terms which would never be used to describe such a stimulus no matter how its spectral composition varied. One such term is grey: a white light seen in a dark-room can be reduced in intensity to the point of imperceptibility without an observer describing its colour as grey. The sensation of grey is therefore confined to related colours and it is only by surrounding a spot of white light with a brighter white background that it can be made to appear grey. If the background brightness is increased sufficiently it will even appear black. This effect can be readily seen on a television screen. When the set is switched off and is viewed under ordinary room lighting it appears light grey or pale green. When it is switched on black objects appear black even though the amount of light reflected from the black area is unchanged. This situation, of course, involves related colours.

The conditions for perceiving a sensation of brown are exactly the same and the reader might like to predict what coloured light will appear brown if seen against a much brighter white background. Presumably observers who find it impossible to describe the perceived hue of a brown in terms of the red, yellow, green and blue psychological primaries will not be able to do this. Some of those who took part in the Padgham and Rowe experiment estimated that the hue appeared to consist of redness and blueness with the latter predominating: these observers would be expected to answer 'purple'. However, the light which is best able to evoke the sensation of brown when surrounded by a much brighter white is invariably described as orange when seen against a dark background. All the colours a normal observer would describe as brown can be matched by superimposing a neutral grey filter over one of the orange specimens in the *ICI Colour Atlas*. A comprehensive study into the perception of brown has been made by Bartleson (1976).

The dyer's variables of perceived colour

The dyer also recognises three independent variables of perceived colour but only one, that of hue, is based purely on observation. As mentioned in chapter 4, dyers describe hue differences in terms of the comparatives of the four psychological primaries – redder, yellower, greener, bluer with one important restriction. This arises if a dyeing has the colour of one of the psychological primaries in which case the associated comparative would never be used, e.g. a red would never be described as redder than another red. The reason for this is simple: the term is ambiguous. The dyer also recognises the validity of the opponent-colours theory in that he would never describe a red as greener than another red, for example.

The second variable is used to describe the differences in appearance caused by differences in the amount of a single dye on the fibre or of a mixture of compatible dyes. This variable is termed 'depth', the comparatives being 'fuller' and 'thinner' (Morton 1976). Increases in the amount of dye on the fibre cause visual effects which have to be learned and which are not simply related to the natural variables of lightness and chroma. For example, increasing the depth of a dyeing made with a neutral grey dye only reduces the lightness whereas in the case of a bright yellow dye, only an increase in chroma occurs.

The fastness properties of a dye depend on the depth and to ensure that published fastness ratings are comparable ISO issues a set of standard depths in eighteen hues together with a number of supplementary standards as described on p. 62. These were originally chosen by German and Swiss dyemakers in the mid-1930s and the basis of the choice is not known but it must have been intuitive as it is not possible to define any criteria by which the depth of a yellow dyeing can be equated with that of a blue dyeing, for example.

If two dyeings have the same hue and the same depth they will not necessarily match and the difference between them constitutes the dyer's third variable – 'brightness', the comparatives being 'brighter' and 'flatter' (Morton 1976). Differences in brightness can be eliminated by adding the correct amount of a neutral grey dye to the brighter specimen, this having the effect of decreasing both the lightness and the chroma.

The relationship between the dyer's variables and the Munsell variables will be considered in detail in chapter 11.

8

The Measurement of Colour

The basis of the modern method of colour measurement was outlined in chapter 4 as it emerged from the fundamental studies of colour vision made by Helmholtz, Grassmann and, in particular, Maxwell. If we have a device which enables us to mix monochromatic red, green and violet lights in any proportion, all the colours which can be obtained with dyes and pigments can be matched perfectly. Such a match can be represented as a colour matching equation

$$Q \equiv R + G + V$$

the amounts of each of the four components being implied but not designated for simplicity.

The colours of other monochromatic lights, however, cannot be matched but each can be incorporated into a colour matching equation if one of the primaries is added to it, for example

$$Q + R \equiv G + V.$$

Maxwell's vital contribution to colour measurement was to realise that in this case the amount of R which has to be added to Q so that the mixture can be matched by $G + V$ can be regarded as a negative amount, namely

$$Q \equiv G + V - R.$$

Maxwell also envisaged the possibility that two of the primaries might have to be added to Q, for example (Maxwell 1860)

$$Q + R + G \equiv V$$

therefore

$$Q \equiv V - R - G.$$

This approach meant that it was unnecessary to restrict the primaries to red, green and violet: colour matching equations could be set up with any three reference lights providing none could be matched by an additive mixture of the other two. The only benefit of this extension, however, was the use of a blue primary rather than violet as this required much less energy to achieve the same brightness as shown by the curve showing the relative brightness of equal-energy amounts of lights at each wavelength (figure 8.5).

If Q is matched by specific amounts of $R + G + B$, any other stimulus which matches Q, irrespective of its spectral composition, will also be matched by the same amounts of R, G and B. These amounts therefore quantify the colour of Q irrespective of its composition.

Colour matching equations behave as though they were algebraic equations. For example, the light obtained by passing the light from a tungsten filament lamp through a yellow gelatin filter can be perfectly matched by an additive mixture of monochromatic red light and green light

$$Y \equiv R + G.$$

If equal amounts of another light S are added to both sides of the equation the match will still hold even though the colour may be entirely different, i.e.

$$S + Y \equiv R + G + S.$$

If S is then matched by another light T, the following equations are valid

$$S + Y \equiv R + G + T$$

and

$$T + Y \equiv R + G + S.$$

Finally, if the amount of each of the components of a colour matching equation is multiplied by the same factor, the match will still hold providing that the factor is not so small that the rods in the retina play a significant part of the visual process nor so large as to cause the eye to be dazzled.

From these facts concerning additive colour mixing Maxwell defined an objective method of colour measurement. Although this method was based on the trichromatic theory of colour vision its validity does not, in any way, depend on the correctness of that theory. This is used today and, given the present state of colour vision theory, it is not possible to conceive of any other method. Maxwell stated that it was first necessary to specify completely these additive primaries and that the most satisfactory would be monochromatic red, green and blue lights. The amounts of these primaries which are required to establish a colour matching equation with monochromatic lights throughout the visible spectrum must then be determined using an instrument presenting a split field to the eye; one half consists of monochromatic light of known wavelength, the other half an additive mixture of the three primaries. Provision must also be made, of course, to add one of the primaries to the monochromatic light being matched. In the literature these amounts have been given various names – excitation curves, distribution coefficients, spectral tristimulus values – but the most appropriate is *colour matching functions*, the term currently favoured by the CIE. Once they have been determined the algebraic nature of colour matching equations then permits the calculation of the amounts of the primaries required to match *any* stimulus whose spectral power distribution is known. These amounts therefore define the colour of any stimulus irrespective of its spectral composition and are termed *tristimulus values*.

The determination of the colour matching functions

In 1858 Maxwell built an apparatus for determining colour matching functions which he called a 'colour box', the modern term being 'visual additive colorimeter'. Maxwell's wife Kathleen acted as his observer. Subsequent measurements were made with ever increasing precision by König and Dieterici (1892), by Abney (1913) and independently by Guild (1931) and Wright (1928, 1929) in the late 1920s. These last two studies were made using markedly dissimilar apparatus: Guild used heterochromatic primaries obtained with coloured filters whereas Wright used monochromatic primaries at 650, 530 and 460 nm. There were significant differences in the calibration procedures employed as well but the algebraic nature of colour matching equations meant that the results could be converted to the same set of primaries, 700, 546.1 and 435.8 nm, the last two being the wavelengths of prominent mercury lines and therefore easily obtained by filtering the light from a mercury vapour lamp. These primaries were normalised so that equal amounts were required to match a stimulus containing equal energy amounts of light at each wavelength in the visible region. When this was done the colour matching functions of the seventeen observers were in such good agreement that the CIE, in 1931, could take the means as defining the characteristics of an average observer at 5 nm intervals from 380 to 780 nm.

The method the CIE devised for using these mean values to define a colour numerically has been described by one of the pioneers of colour science, W A H Rushton (1975), as being

> like the square root of −1 in alternating-current theory [it] gives neat and exact answers to calculations by introducing a nearly incomprehensible concept of what is really going on.

Readers interested in the experimental details of this important development could not do better than to read the definitive account of Wright (1969) whilst those interested in its historical development and in getting a first-hand account of the 1931 CIE meeting are referred to Wright (1970, 1981, 1982). The following, much shortened account, however, may be adequate.

The concept of imaginary primaries

Whilst the mean Guild/Wright colour matching functions could have been used in routine calculations, at every wavelength except those of the three primaries, one function is negative. The CIE thought that calculations requiring the summation of positive and negative numbers would be subject to errors and moreover the mechanical calculators which had already been built into a recording spectrophotometer (see chapter 9) would be useless. They therefore decided that routine calculations should be carried out using colour matching functions which were never negative. This could be done by converting the *RGB* functions into those referring to three *imaginary* primaries. Mathematically, this was no different from converting Guild's and Wright's functions into the same *RGB* functions but the actual choice of imaginary primaries was very important.

If two stimuli are generated from red, green and blue primaries in which the amounts of each differ by the same factor, i.e.

Stimulus 1 = *R* units of red, *G* units of green, *B* units of blue

Stimulus 2 = *nR* units of red, *nG* units of green, *nB* units of blue

the untrained observer would describe the two stimuli as having the same colour but differing in brightness. One familiar with the modern terminology of colour science would also say they differed in brightness but had the same hue and the same saturation or the same *chromaticness*. One therefore requires a set of three values which will be the same for both stimuli and the obvious set is the *ratios* of the amounts of the three primaries. These are termed *chromaticity co-ordinates* and are given by

$$r = \frac{R}{R+G+B} \qquad g = \frac{G}{R+G+B} \qquad b = \frac{B}{R+G+B}$$

R, *G* and *B* being the tristimulus values of the stimuli. Historically, Guild and Wright determined the chromaticity co-ordinates of the spectral colours: in those days it was not possible to measure absolute amounts of the primaries accurately enough but the ratios could be measured to better than three decimal places.

In 1922 the Optical Society of America devised a set of colour matching functions based on the measurements made by König and Dieterici and by Abney. Troland, chairman of their Colorimetry Committee, put forward an extremely valuable method of plotting chromaticity co-ordinates. Up to this time they had been plotted as trilinear co-ordinates based on an equilateral triangle whose apexes indicated the position of the primaries, a method devised by Maxwell for calculating the position of light mixtures by the centre-of-gravity method. Troland pointed out that it would be very much easier to base the method of plotting on a right-angled triangle using only two co-ordinates as ordinary graph paper could then be used (Troland 1922), the third co-ordinate being redundant as $r + g + b = 1$.

The chromaticity co-ordinates of the spectrum corresponding to the *RGB* reference primaries chosen by the CIE are shown in figure 8.1. The maximum negative value reached by any co-ordinate is very large solely because of the wavelength chosen for the green primary: had this been 510 nm the maximum value would have been much smaller.

Now the position of the spectrum can be specified with reference to any other triangle drawn in this diagram and if such a triangle completely encloses the spectrum locus, no chromaticity co-ordinate and therefore no colour matching function will be negative. The apexes of this new triangle will locate the position of the new primaries and as at least two must lie outside the spectrum locus they will be imaginary. The triangle should clearly be as close to the spectrum locus as possible and it is obvious that this could be achieved in the

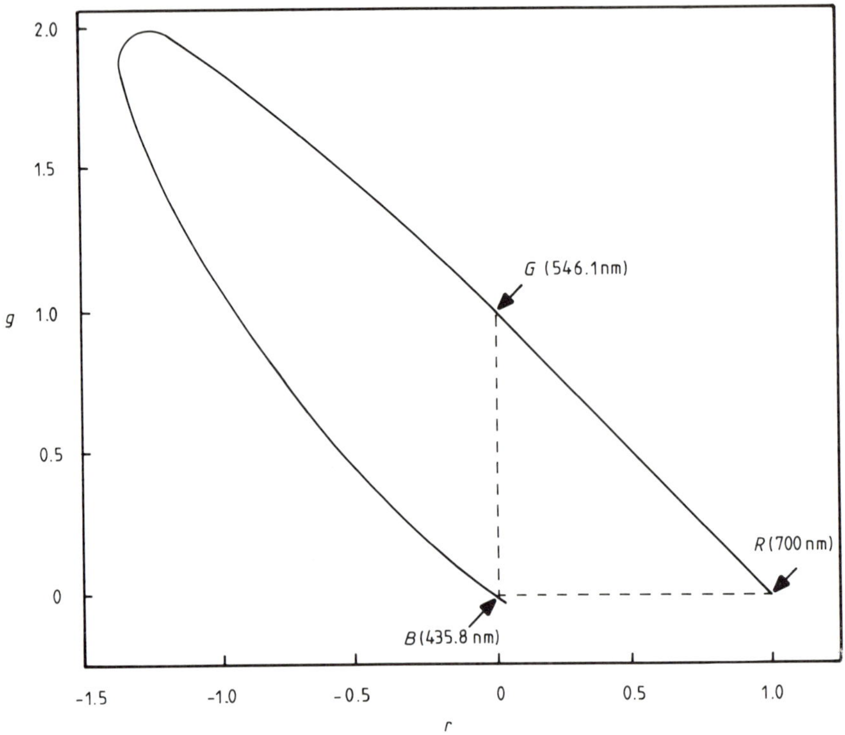

Figure 8.1 Chromaticity co-ordinates of the spectrum in the *rg* diagram.

manner described. Zero amounts of the blue primary are required to match the spectrum beyond 660 nm and therefore the locus is a straight line from 660 to 780 nm. This line, extrapolated from 780 beyond 660 nm, could form the first side: it would have the advantage of eliminating the *B* calculations for all wavelengths longer than 655 nm. The second side could then be the straight line joining the ends of the spectrum, the loci of mixtures of monochromatic red and violet lights, the purple line. This side would then have to be extrapolated and the third side could then be drawn tangential to the spectrum locus.

The CIE adopted the position of the first side, but rejected that of the second in favour of a device originally suggested by Schrödinger (1920) and elaborated in a very impressive paper by Judd (1930). It is possible to locate this side so that the imaginary primaries at both ends have zero luminance. This procedure has the extremely valuable consequence that the luminance of any stimulus is then quantified by the tristimulus value referring to the third primary. How this was achieved can best be understood by a geometric approach devised by Blakey and Landon (1978).

Figure 8.1 is a two-dimensional diagram quantifying hue and saturation and in figure 8.2, the third dimension, luminance, is introduced orthogonal to the

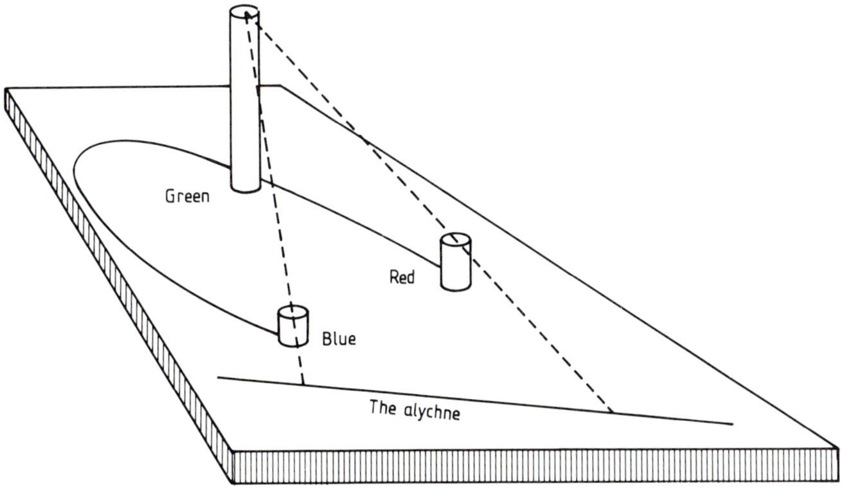

Figure 8.2 The derivation of the alychne (see text).

chromaticity plane. The relative luminances of the amounts of the primaries required to match the equal-energy spectrum were 1 (red), 4.4907 (green) and 0.0601 (blue). These are shown by the heights of the columns, but not to scale for clarity. Luminances are additive so that the line joining the tops of the two columns quantifies the luminance of equal-energy amounts of mixtures of the two lights. If the line joining the top of the 'green' column and the tops of the other two are then extrapolated to cut the plane of the chromaticity diagram, the line joining these two points is termed the *alychne*. Points lying on this line represent imaginary lights because it lies wholly outside the spectrum locus and these imaginary lights, also have zero luminance. If two of the primaries are located on the alychne then the whole of the brightness of any stimulus is quantified by the tristimulus value corresponding to the remaining primary. The importance of this cannot be exaggerated. The first side cuts the alychne at X in figure 8.3 and the third side of the triangle was located to pass near the spectrum locus more or less arbitrarily and its intersection with the other two sides located the other two imaginary primaries Y and Z.

The CIE then calculated the colour matching functions of the standard observer in terms of these XYZ primaries at every 5 nm from 380 to 760 nm to five places of decimals.. These have now been interpolated and extrapolated to every one nanometre from 360 to 830 nm to twelve places of decimals although virtually all industrial requirements are met by calculation at 20 nm intervals from 400 to 700 nm to four places.

The composition of the stimulus reflected from a surface depends not only on its reflectance at each wavelength but also on the composition of the light falling on it. The CIE therefore defined three illuminants A, B and C representative of the light from a tungsten filament lamp, direct sunlight and overcast

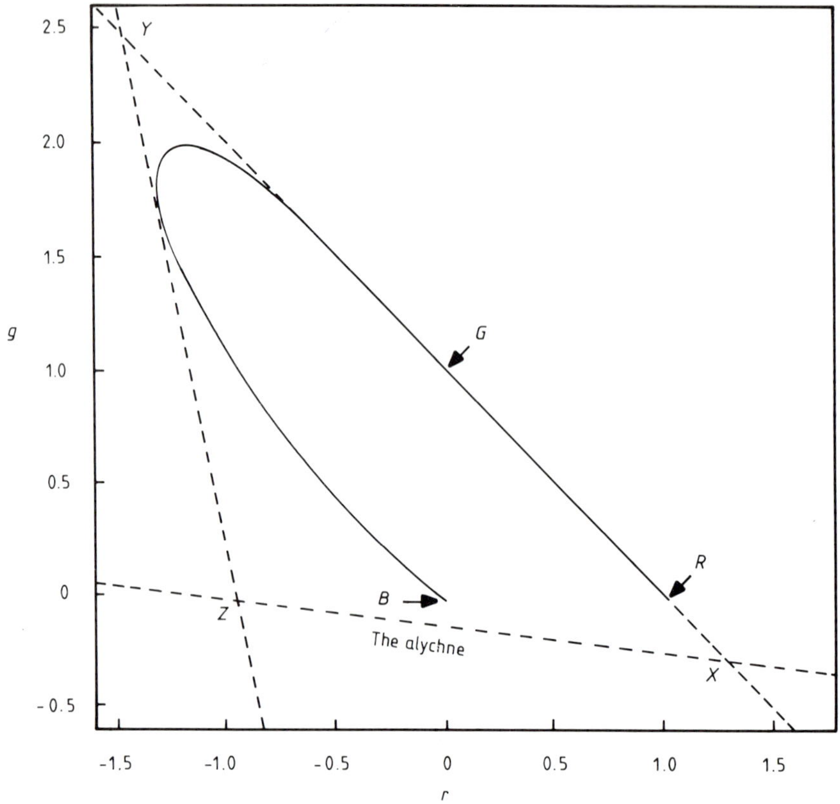

Figure 8.3 The origin of the *XYZ* imaginary primaries (see text).

daylight. These were defined in two different ways which reflected divergent views as to how standards should be defined. Guild was adamant that a source should be defined in such a way that it could be set up in any standards laboratory: this would mean defining filters by their chemical composition. He also believed that routine colour measurement would be carried out using visual colorimeters which would also require realisable light sources. Priest, for the USA, however, was equally convinced that colour measurements would be made by calculation from reflectance curves and therefore that illuminants would have to be defined by their spectral power distributions. The inevitable compromise occurred – illuminants A, B and C were defined in both ways and later the CIE differentiated between them by referring to the lamp/filter combinations as sources and the SPD as illuminants. The American view as to how the nascent technology of colour measurement would be carried out prevailed and sources A, B and C were rarely set up which is fortunate as it has now been found that when they are, sources B and C do not have the SPD of illuminants B and C (Billmeyer and Alessi 1981). Illuminant B was rarely used: the SPD of illuminants A and C are shown in figure 8.4.

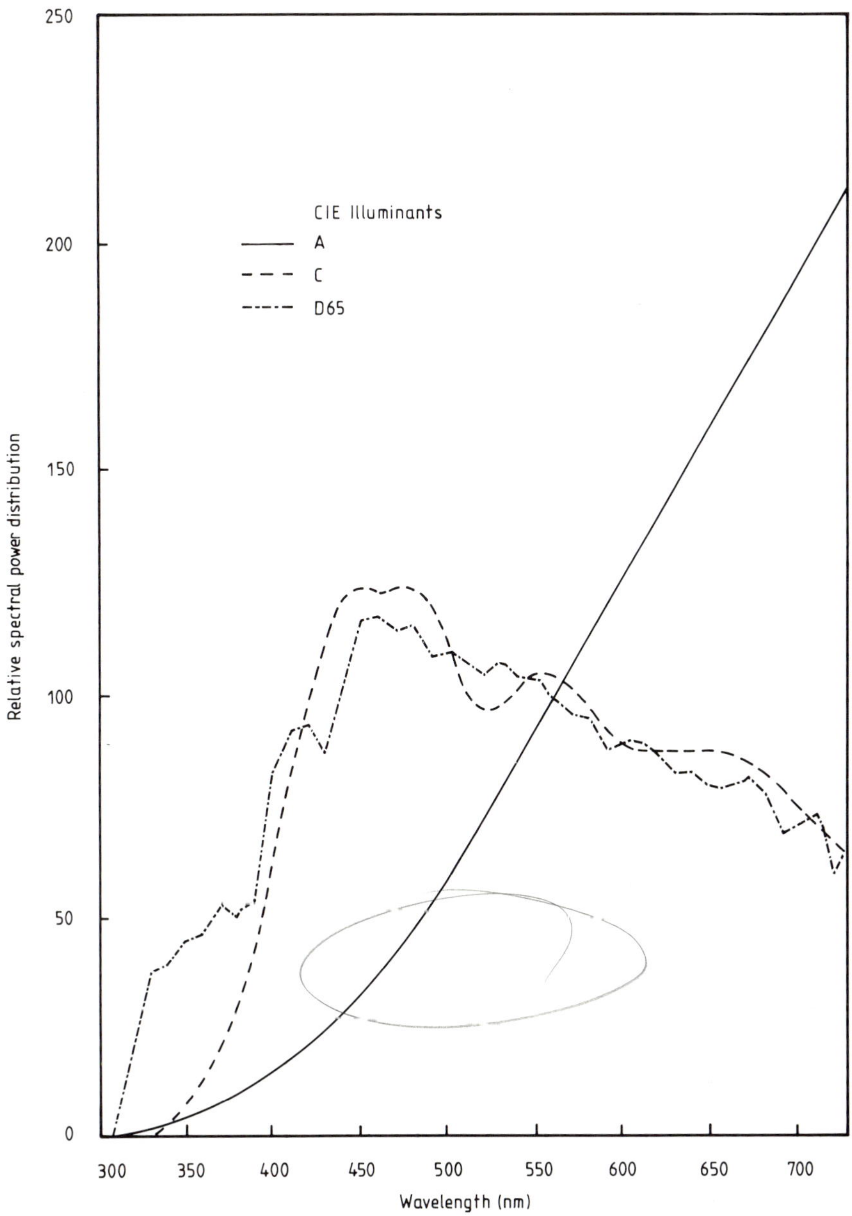

Figure 8.4 Spectral power distributions of CIE illuminants.

The calculation of tristimulus values

Tristimulus values are calculated from spectral reflectance factors, $R(\lambda)$, covering at least the range 400–700 nm. These factors express on a 0–1 scale the reflectance at each wavelength as a decimal fraction of that reflected by a perfect reflecting diffuser identically illuminated. (*Note:* until 1969 the standard of reflectance factor was freshly smoked magnesium oxide.) At each wavelength interval, $R(\lambda)$ is multiplied by $S(\lambda)$, the relative spectral power distribution of the illuminant and by each of the three colour matching functions $\bar{x}(\lambda)$, $\bar{y}(\lambda)$ and $\bar{z}(\lambda)$. Three summations are then made over the chosen range

$$X = k \sum R(\lambda)\, S(\lambda)\, \bar{x}(\lambda)$$

$$Y = k \sum R(\lambda)\, S(\lambda)\, \bar{y}(\lambda)$$

$$Z = k \sum R(\lambda)\, S(\lambda)\, \bar{z}(\lambda)$$

The scaling factor k has to be introduced because otherwise the summations would depend on the wavelength interval chosen. This is given by

$$k = 100 \left(\sum S(\lambda)\, \bar{y}(\lambda) \right)^{-1}$$

which means that the Y value of the perfect reflecting diffuser and therefore of the illuminant used in the calculations is always 100.0.

The 10° observer data

The angle subtended by the split field in Guild's and Wright's colorimeters and that used in the determination of the $V(\lambda)$ curve was 2°, the diameter of the field being about the width of the thumb held at arm's length. This was considered desirable because of a phenomenon caused by macular pigmentation (see chapter 6). Any stimulus of extreme spectral energy distribution, such as mixtures of monochromatic primaries, which fills a field of at least 10° appears non-uniform. The non-uniformity consists of a spot, the so-called Maxwell spot, with an ill-defined boundary and a diameter of about 4° which differs in colour from the rest of the field and which does not remain in the same place. A field size of 2°, however, appears perfectly uniform. As the Maxwell spot is caused by macular pigmentation it follows that strongly metameric matches (see chapter 6) such as those made in a visual additive colorimeter will be different for the major part of a large field than for the 4° centre. As industrial colour matching invariably involves field sizes much greater than 2° the CIE, in 1964, published colour matching functions for a 10° observer. This was based on experimental work carried out by Stiles and Burch (1959) and by Speranskaya (1959) in which a total of 67 observers matched monochromatic lights from approximately 390 to 830 nm with additive mixtures of monochromatic lights subtending a 10° field in which the centre 4° region was either ignored or masked off. The CIE transformed these mean data to derive $\bar{x}_{10}(\lambda)$, $\bar{y}_{10}(\lambda)$ and $\bar{z}_{10}(\lambda)$ colour matching functions, the resultant Y_{10} tristimulus value alone quantifying brightness but not luminance as this term today is confined to calculations made using the 2° data.

For most industrial purposes the 10° observer data is preferable and not only because visual matching is normally carried out with specimens subtending an angle greater than 10° to the eye. The 2° *Y* value is a measure of luminance and in 1924 the CIE had developed a method of quantifying luminance based on the relative brightness of equal energy amounts of light of each wavelength, the $V(\lambda)$ curve, figure 8.5. It was obviously essential that both measures should give exactly the same values but in order to do this 'some fiddling had to be done with the actual relative luminances of the *RGB* stimuli' (Wright 1970). In 1949, however, it was realised that the early data were excessively low at the violet end of the spectrum by a factor of over ten times at 400 nm (Judd 1949, 1950). For various reasons, mainly practical, the CIE decided not to change either $V(\lambda)$ or $\bar{y}(\lambda)$ and as the 10° data are free from this error this is another reason why they should normally be used. The only exceptions are when critical colour areas subtended an angle of less than 4° to the eye, for example, high visibility clothing, road signs and colour photography: in such cases the 2° observer data have been found to correlate better with visual observations where metamerism is involved in spite of its short wavelength error.

Illuminant D65

Illuminant C was based on measurements of daylight in the visible region and was completely satisfactory until fluorescent brightening agents became industrially important in the late 1940s. Then the fact that it was significantly

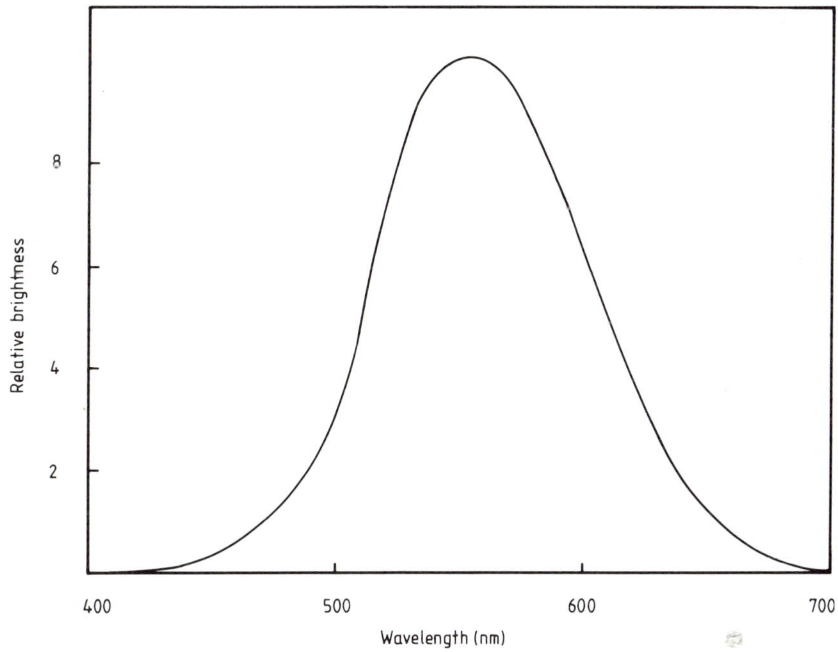

Figure 8.5 Relative brightness of equal-energy amounts of lights of each wavelength between 400 and 700 nm.

lower in the ultraviolet region than daylight resulted in it being augmented by a series of D illuminants. Measurements over the range 330–700 nm made at Enfield (Middlesex), Rochester (New York) and Ottawa showed that the spectral composition of daylight depended on its colour temperature, the place being unimportant (Judd *et al* 1964). From the mass of data obtained, SPD extrapolated to 300 and 830 nm were calculated for several colour temperatures of which the most important was that at approximately 6500 K, the exact value depending on the value adopted for the constant c^2 in Planck's radiation law. Its SPD is shown in figure 8.4.

Unfortunately the CIE could not adopt the dual approach of 1931 when a physically realisable source and a table of data were established. Only the latter was possible in 1966 and the CIE differentiated between the two by using the term *illuminant* for data and *source* for a realisable light: source C and illuminant C both exist but D65 is only an illuminant. It is ironic that whereas source C was available but rarely needed, source D65 is necessary but seems unlikely ever to be realised in a form suitable for incorporating in a spectrophotometer and the problems this has caused will be discussed in the next chapter.

Three-band lamps

Although fluorescent lamps were introduced just before the Second World War, they were rarely used as illuminants in colorimetry mainly because their colour rendering properties left so much to be desired that objective quantification was unnecessary. The exception was the lamp covered by BS 950: Part 1 which simulated daylight closely enough to suffice for most purposes. In 1975, however, a new generation of fluorescent lamps arrived based on the rare-earth phosphors used in colour television. These lamps were characterised by a high proportion of the emitted light being in three narrow bands in the red, green and blue spectral regions, figure 8.6. Such a spectral power distribution gave a luminous efficiency as high as the most efficient fluorescent lamps but with very much better colour rendering though not as good as that given by lamps conforming to BS 950: Part 1. Three-band lamps, however, had a marked effect on many metameric matches which were acceptably close under conventional sources as we shall see in chapter 10.

The *xy* chromaticity diagram

The colour matching functions, $\bar{x}(\lambda)$, $\bar{y}(\lambda)$ and $\bar{z}(\lambda)$, are also the tristimulus values of the equal-energy spectrum and therefore at each wavelength the chromaticity co-ordinates are given by

$$x = \frac{\bar{x}}{\bar{x}+\bar{y}+\bar{z}} \qquad y = \frac{\bar{y}}{\bar{x}+\bar{y}+\bar{z}} \qquad z = \frac{\bar{z}}{\bar{x}+\bar{y}+\bar{z}}.$$

These are not merely the co-ordinates of the equal-energy spectrum: as they are ratios they apply to any energy value. They are, of course, always plotted in a right-angled triangle and, given the mathematical orientation of the founders of

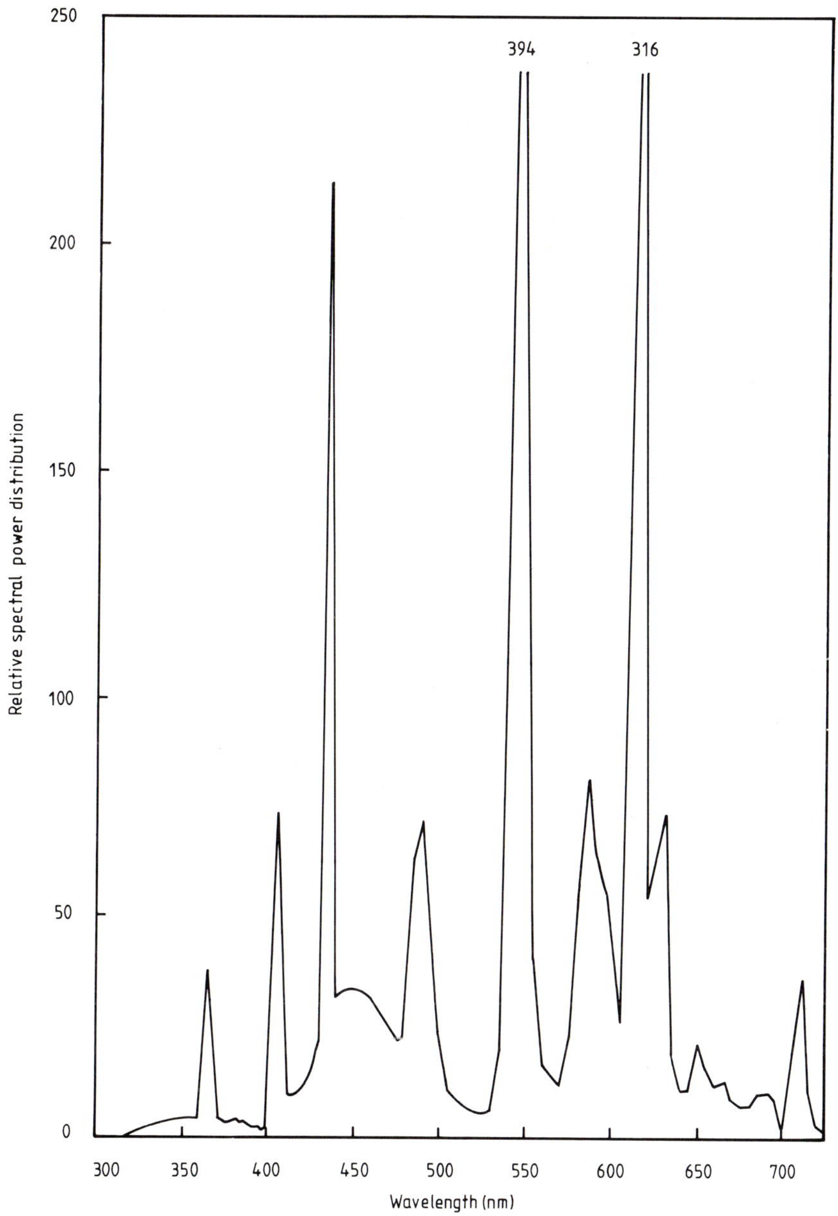

Figure 8.6. Spectral power distribution of a typical three-band fluorescent lamp (Philips Colour 84).

the CIE system, it was inevitable that x should be plotted against y though there is no colorimetric reason for so doing. Uyterhoven (1938) plotted y against z and Adams (1942) favoured x against z but the xy diagram became universally adopted, the hypotenuse being omitted for clarity. The 10° observer diagram is shown in figure 8.7, that for the 2° observer being very similar in shape but with marked displacements of certain wavelengths.

From the locations of two stimuli, m and n, two variables of perceived colour can be quantified. The wavelength where the line from D65 through m cuts the spectrum locus (550 nm) is the *dominant wavelength* (λ_d) of m. The wavelength where the extrapolated line from D65 through n cuts the spectrum locus (490 nm) is the *complementary wavelength* (λ_c) of n. The *excitation purity* of m and n is the fraction of the distance out towards the spectrum locus/

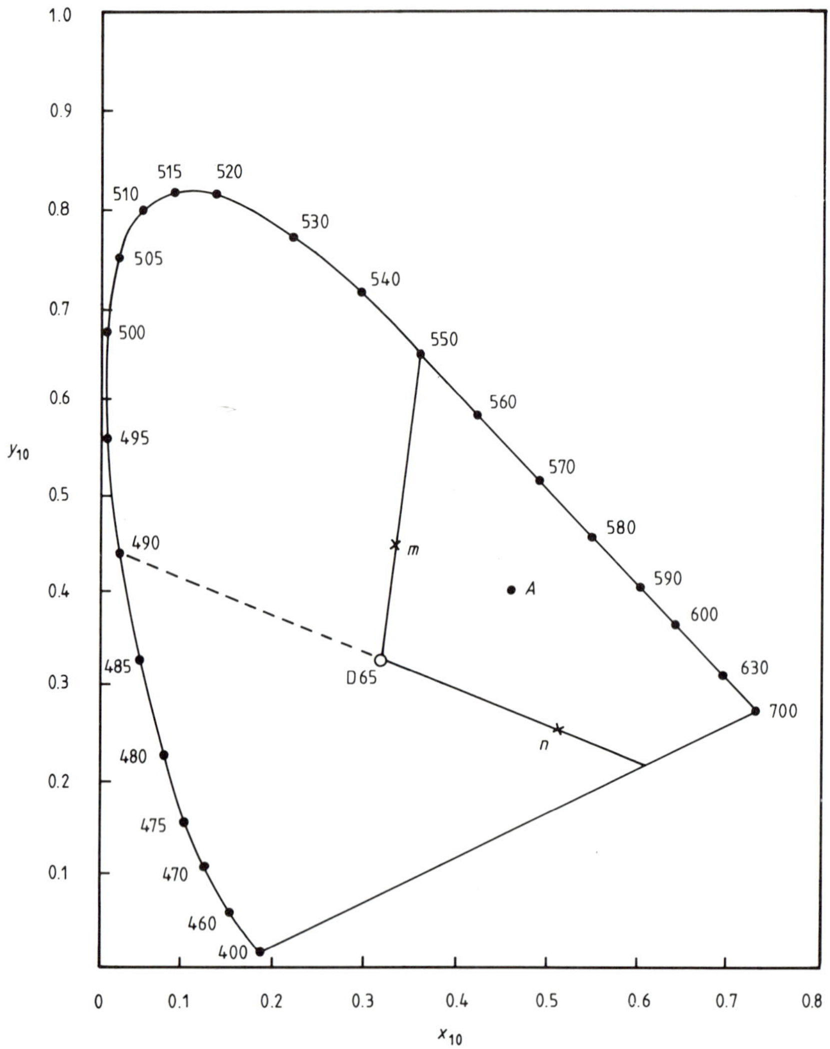

Figure 8.7 x_{10}, y_{10} chromaticity diagram.

purple line from D65. As the *Y* value quantifies the lightness of a surface colour, *xy Y* values are therefore much more suitable than *XYZ* values for quantifying appearance and, in consequence, they were more widely used. They are today, however, being replaced by CIELAB values which are better still as we shall see in chapter 11.

9

Colour Measuring Instruments

9.1 REFLECTANCE SPECTROPHOTOMETERS

A reflectance spectrophotometer is basically an instrument for measuring the amount of light reflected from the surface of an opaque specimen at a number of wavelengths throughout the visible spectrum as a fraction of that reflected by a white standard identically illuminated. This fraction is termed the *reflectance factor* (CIE definition). For many years the white standard was a block of magnesium carbonate on which had been deposited a layer of magnesium oxide obtained by burning magnesium ribbon. This was assumed to reflect all the incident light but was subsequently found to reflect only between 97 and 98%: moreover it was difficult to prepare consistently and had only a limited life. The CIE therefore recommended that from January 1, 1969 the absolute standard should be the perfect reflecting diffuser whose reflectance at every wavelength is 100%. Such a surface does not exist, of course, but working standards chosen for stability and ease of use can be calibrated in terms of the perfect reflecting diffuser.

If a beam of light is shone on such a surface the reflectance factor at every angle is the same. This is represented for a beam at 45° by the goniophotometric curve shown in figure 9.1(*a*). At the other extreme a perfect mirror surface would reflect all the light at −45°. The surfaces of objects coloured with dyes or pigments usually fall between these two extremes and if there is a significant specular (often termed 'regular') component then the illuminating and viewing conditions are critically important. Goniophotometric curves of two such surfaces are shown in figures 9.1(*b*) and (*c*), the radial distance of the curve from the centre of the specimen being proportional to the reflectance factor measured in that direction. The illuminating and viewing conditions are important because, even if the object is highly coloured, if specular reflection occurs it occurs at the surface of the medium before selective absorption by the colorant can occur – highlights are therefore white unlike those from a metallic surface which have the same colour as the metal itself.

The fraction of the incident light which is specularly reflected at the surface of a medium whose refractive index is n is given by the Fresnel relationship

$$(n-1)^2/(n+1)^2.$$

Plate 1 The Munsell Color Tree.

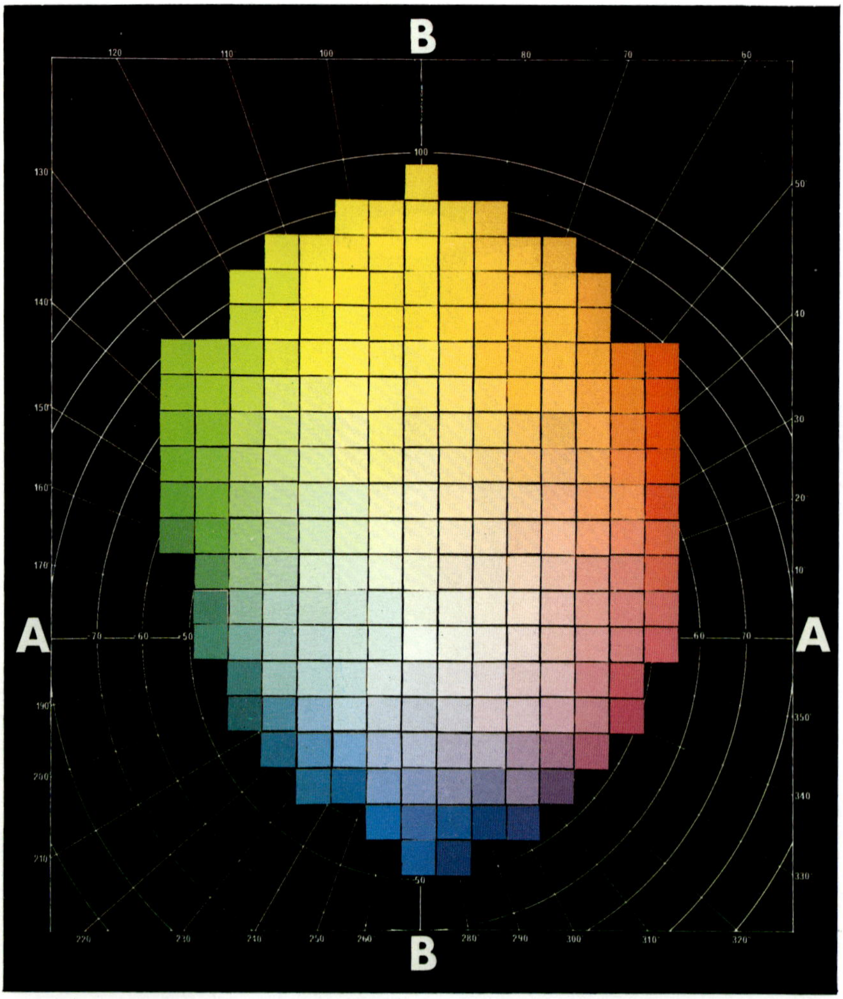

Plate 2 The CIELAB a^*b^* diagram.

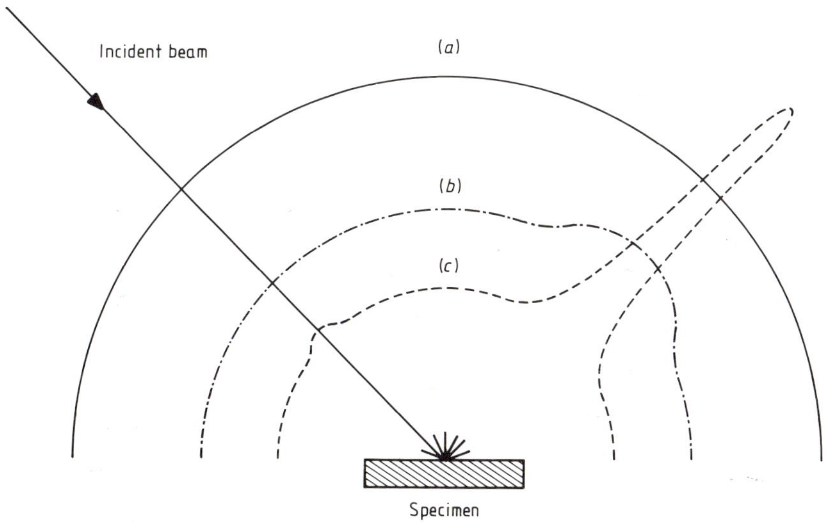

Figure 9.1 Goniophotometric curves of (*a*) the perfect reflecting diffuser; (*b*) egg-shell finish paint; (*c*) high-gloss paint.

The refractive index of a typical paint medium and of most plastics etc is about 1.54 so that if the specular component is included during measurement the reflectance at each wavelength will be raised by about 4.5% above that obtained when the specular component is excluded. The reflectance at the blue end of the spectrum of a high gloss yellow paint will therefore be raised from, say 1 to 5.5%, that at the red end from, say 80 to 84.5%.

For colorimetry the CIE has defined four different illuminating and viewing conditions.

(*a*) *45°/normal (abbreviation, 45/0).* The specimen is illuminated by one or more beams whose effective axes are at an angle of 45 ± 2° from the normal to the specimen surface. The angle between the direction of viewing and the normal to the specimen should not exceed 10°. The angle between the axis and any ray of an illuminating beam should not exceed 8°. The same restriction should be observed in the viewing beam.

(*b*) *Normal/45° (abbreviation, 0/45).* The specimen is illuminated by a beam whose effective axis is at an angle not exceeding 10° from the normal to the specimen. The specimen is viewed at an angle of 45 ± 2° from the normal. The angle between the axis and any ray of the illuminating beam should not exceed 8°. The same restriction should be observed in the viewing beam.

(*c*) *Total/normal (abbreviation, t/0): diffuse/normal (abbreviation, d/0).* The specimen is illuminated by an integrating sphere. The angle between the normal to the specimen and the axis of the viewing beam should not exceed 10°. The integrating sphere may be of any diameter provided the total area of the ports does not exceed 10% of the internal reflecting sphere area. The angle between the axis and any ray of the viewing beam should not exceed 5°. If all regularly reflected light is included in the measurements, the condition is t/0; if all regularly reflected light is excluded the condition is d/0.

(*d*) *Normal/total (abbreviation, 0/t): normal/diffuse (abbreviation, 0/d).* The specimen is illuminated by a beam whose axis is at an angle not exceeding 10° from the normal to the specimen. The reflected flux is collected by means of an integrating sphere. The angle between the axis and any ray of the illuminating beam should not exceed 5°. The integrating sphere may be of any diameter provided the total area of the ports does not exceed 10% of the internal reflecting sphere area. If all regularly reflected light is included in the measurements, the condition is 0/t; if all regularly reflected light is excluded the condition is 0/d.

The reflectance spectrophotometers most widely used in industry have the t/0 and d/0 geometries as a light trap can be inserted in the sphere wall to exclude the regularly reflected light. These options are normally designated 'specular included' (SPINC) or 'specular excluded' (SPEX).

Though it would appear that there are six different illuminating and viewing geometries, and spectrophotometers based on most of them are commercially available, there are, in effect, only three. This is because the Helmholtz reciprocity principle states that the reflectance is the same if the directions of illumination and view are interchanged. Though the values obtained with any of these viewing geometries are spectral reflectance factors as defined by the CIE, they are invariably called *reflectances* especially today as a modern spectrophotometer is usually associated with a visual display unit of limited area. The CIE definition of *reflectance*, however, is the ratio of reflected to incident radiant flux and is given the symbol ρ. Thus only the values obtained using a 0/d SPINC geometry are reflectance values even though identical values are given by a d/0 SPINC geometry. Reflectances are invariably expressed as percentages for convenience but it is essential to remember that they must be divided by 100 before beginning *any* colorimetric calculation.

The CIE also recommends that the value obtained whenever the angle of view approaches 0° should be termed a *spectral radiance factor* and given the symbol β in accordance with the radiometric quantity *radiance* but this would seem to be unnecessary when objects are not fluorescent. When they are, however, the term reflectance should not be used because in one region of the spectrum only part of the light coming from the specimen has been reflected while part has been emitted as a result of the fluorescence mechanism. The values obtained for fluorescent specimens should therefore be termed *spectral radiance factors* and subdivided where appropriate into a *reflected radiance factor* and a *fluorescent radiance factor*. If the measurements have been made with a spectrophotometer based on a 0/d geometry, however, the values are neither spectral reflectance factors nor spectral radiance factors according to the strict CIE definitions: fluorescence prevents all of them being reflectance factors and none are radiance factors because the angle of view is 2π steradians instead of 0°. The only permitted term would appear to be *reflectometer values* which the CIE recommends for any values obtained from an instrument measuring reflectance factors. However, this term does not seem to have often been used.

The choice of viewing geometry

The effects of the three basic viewing geometries on the spectral reflectance factors of specimens which are not matte can be quite marked as is shown by the tristimulus values and chromaticity co-ordinates of the yellow standard ceramic tile (see later) which are given in table 9.1.

Table 9.1 Tristimulus values and chromaticity co-ordinates (2°/D65) of a glossy yellow tile measured using three different viewing geometries.

Geometry	X	Y	Z	x	y
0/45	57.19	62.65	12.48	0.4322	0.4735
0/d (SPEX)	55.50	60.75	12.19	0.4321	0.4730
0/d (SPINC)	59.44	64.91	16.88	0.4209	0.4596

It is interesting to note that the tristimulus values are not lowest for the 0/45 geometry as one would expect. An efficient gloss trap appears to exclude more of the specular component than the 0/45 geometry.

As professional colourists usually try to prevent the specular component entering the eye it is axiomatic that most measurements made with integrating sphere instruments should be made in the SPEX mode but there are two exceptions. Unfortunately the efficiency of the gloss trap varies quite markedly between instruments so that if it is desired to minimise the variations between instruments – in setting colorimetric specifications, for example – SPINC is preferred as this has been found to halve the variations (Billmeyer and Hemmendinger 1981). The second is in computer match prediction where experience has shown that more accurate predictions result if all measurements are made in the SPINC mode and the Saunderson correction is applied as described in chapter 12.

The main advantage of 0/45 and 45/0 geometries lies in on-line colour measurement as these geometries are much less sensitive to slight displacements of the specimen away from the aperture. Orientation sensitivity can be eliminated by complete annular irradiation at 45°. Another advantage claimed for these geometries (Alman and Billmeyer 1976) is that they avoid the variations in spectral radiance factors caused by the paint used to coat the sphere wall having a lower reflectance at shorter wavelengths. Whilst this is completely unimportant if specimens are not fluorescent it can result in the surprising consequence that the factors may be higher for matte specimens if the specular component is excluded. The reason for this is that the reduction in sphere wall area by the gloss trap minimises the error. In practice, however, these effects are trivial compared with those caused by source variations which will be considered later.

The evolution of the recording spectrophotometer

Before 1928 spectrophotometry required the operator to adjust the intensity of monochromatic light falling on a reference white so that it matched that reflected by the specimen being measured, wavelength by wavelength. This was an operation requiring several hours and gave values lacking in precision. It was therefore a major advance when, in that year, the first automatic recording spectrophotometer (figure 9.2) was built by A C Hardy who occupied the Chair of Optics and Photography at the Massachusetts Institute of Technology (Hardy 1929). It employed a beam of white light directed alternately on to the specimen and the reference white, in both cases normal to the surface. This 'double beam' approach has since been characteristic of all precision instruments though it is claimed that modern precision electronic techniques have rendered it unnecessary (Stanziola *et al* 1977). The light reflected at 45° was passed through a single-prism monochromator and a shutter driven by a servo motor reduced the amount of light reflected by the reference white to equal that coming from the specimen as the wavelength of the illumination varied from 400 to 700 nm. The servo motor was coupled to a pen-recorder so that a reflectance curve was plotted, the measurement time being 2½ minutes. 45/0 and 0/45 geometries were consequently specified by the CIE in 1931 but in the following year Hardy redesigned the instrument. Monochromatic light replaced white light because the 2½ minute scanning time caused so much heating of the specimen that the reflectance factors of many were therefore changed. The 0/45 geometry was also extremely sensitive to specimen orientation if it had a surface structure (e.g. most textiles) so this

Figure 9.2 The first recording spectrophotometer.

was replaced by a d/0 geometry (Hardy 1935). This instrument, affectionately known as the 'Hardy', was marketed by the General Electric Company of the USA and for twenty years it was the dominant instrument. Strangely enough, d/0 and 0/d geometries were not recognised by the CIE until 1967.

The return to white light irradiation

The optical path of the 1932 'Hardy' had become a liability in the late 1940s when fluorescent brightening agents (see chapter 2, §2.2) were extensively marketed. In this instrument monochromatic light illuminated the specimen and the amount reflected was measured as a fraction of that coming from the reference white. If a specimen was fluorescent, however, the light from the specimen did not always have the same wavelength as incident light though the recording device assumed that it did. To solve this problem, spectrophotometers had to revert to Hardy's 1928 design of illuminating with white light and passing the reflected/emitted light through an analyser.

Although a new model of the 'Hardy' in which the scanning time was reduced to 1 minute was introduced (Pritchard and Holmwood 1955) this did not include the 'reversed optics' option which was only announced in the 1970s. By this time, however, the dominance of the 'Hardy' had been successfully challenged by the introduction of a number of *abridged* spectrophotometers. In these the specimen was irradiated with white light and the light from the specimen was passed sequentially through between 16 and 38 narrow-band interference filters, an arrangement which considerably reduced the number of moving parts especially as balancing the two beams was done electronically and not by a servo motor. Experience eventually showed that measurement at sixteen wavelengths was adequate (Brockes 1970). This took about 30 seconds, still long enough, however, to affect true values by surface heating (McLaren 1978a).

The current generation of spectrophotometers was introduced independently by three American manufacturers in 1976, and featured advances in electronics and optics but also some unconventional approaches such as interference wedges and reflectance grating monochromators, silicon-photodiode detectors, logarithmic photometric scales and pulsed xenon illumination (Christie and McConnell 1977, Kishner 1977, Billmeyer and Rich 1978, Stanziola *et al* 1979, Rich and Billmeyer 1979). A reasonably detailed comparative evaluation has also appeared (Billmeyer and Alessi 1981). The most important feature of each of these instruments is that the measurement time is only a few seconds so that the surface heating problem has been entirely eliminated.

The problem of sources for the spectrophotometry of fluorescent specimens

When specimens are not fluorescent the SPD of the light source in the spectrophotometer is relatively unimportant providing that there is sufficient at all wavelengths for precise and accurate measurement of black velvet. This is because the ratio of the light reflected from the specimen to that reflected from the white standard is independent of the amount of light at each wavelength. If specimens are fluorescent, however, the total spectral radiance factor at wavelength λ in the emission region is critically dependent not only on the amount of light of wavelength λ but also on that at every wavelength causing excitation. Total spectral radiance factors are therefore only valid for the source used in measurement and as appearance in daylight is usually the most important, the source must resemble daylight as closely as possible, not only in the visible but also in the ultraviolet as well. It will be recalled that illuminant D65 was established because illuminant C was deficient in this region.

Close simulation of illuminant D65 turned out to be very much more difficult than was expected by the CIE when illuminant D65 was established. This has only been possible, for example, by splitting the beam from a xenon arc, applying different filters to the components and recombining (Gundlach 1977) making a source which was totally unsuitable for incorporation in a spectrophotometer for industrial use. The methods used by instrument manufacturers to simulate illuminant D65 have therefore, of necessity, been much less sophisticated. In order to see how consistent they have been, the author measured a white fluorescent plastic specimen and cards printed with six 'Day-Glo' printing inks on five post-1970 spectrophotometers employing simulated D65 sources and the d/0 SPINC geometry. The variations between instruments were determined by selecting the one giving the highest radiance factor, designated *H* (it was not always the same instrument) and calculating the colour difference in CIELAB units between the radiance factors given by *H* and the other four designated *A*, *B*, *C* and *D* randomly. The results are shown in table 9.2 and the spectral radiance curves for one of the specimens in figure 9.3.

Table 9.2 Colour differences caused by instrument variations: for details see text.

	Colour difference (CIELAB units)			
Fluorescent specimen	*H/A*	*H/B*	*H/C*	*H/D*
Ciba-Geigy plastic No. 12	1.1	2.6	2.9	5.4
Lunar yellow	5.4	9.4	4.1	5.9
Arc chrome	4.3	9.2	6.9	9.6
Blaze	6.5	11.2	12.6	10.7
Fire orange	4.9	9.8	7.8	8.3
Neon red	7.6	10.7	10.4	11.3
Magenta	7.2	8.7	9.7	9.0

Although it was not possible to repeat this investigation with non-fluorescent specimens, other much more extensive comparisons (Billmeyer and Hemmendinger 1981) clearly indicate that inter-instrument variations would be less than one-tenth of these. This must mean that the cause lies in the differences in SPD of the light falling on the specimen during measurement.

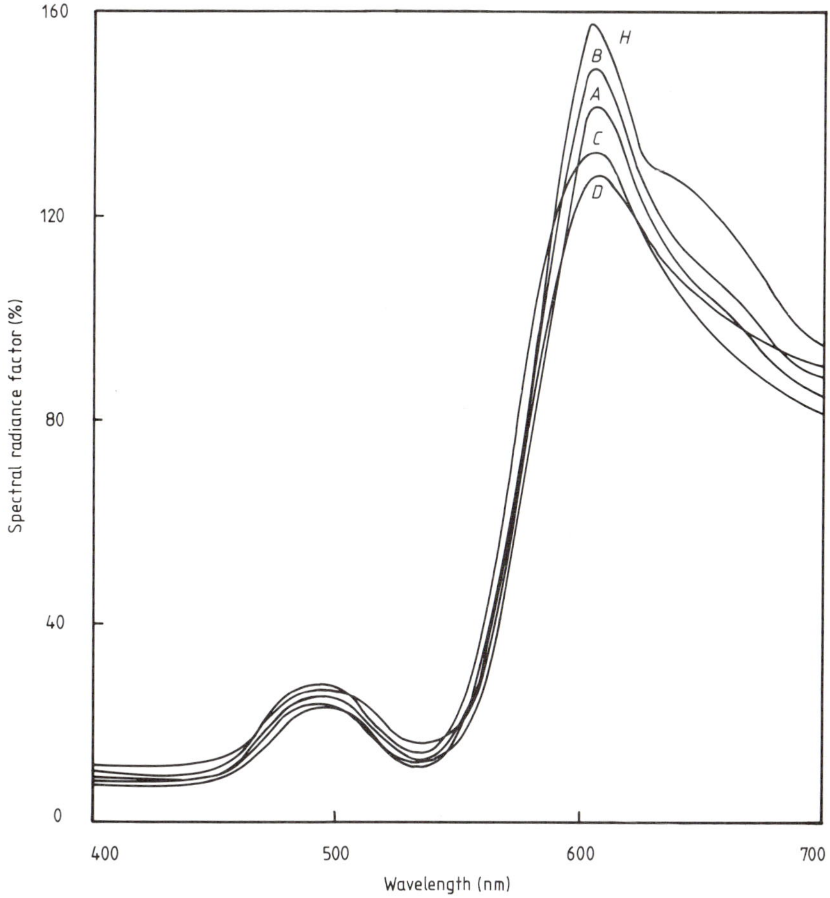

Figure 9.3 Spectral radiance curves of fluorescent orange printing ink measured on five different spectrophotometers using sources simulating illuminant D65.

This lack of precision means that a problem will arise if fluorescent colours are specified by setting permitted areas in the *xy* diagram and minimum *Y* values as was done successfully by the CIE for coloured light signals in the 1930s. The problem is that some specimens may be well within the limits when measured on one commercially available spectrophotometer but well outside on another. The production of a standard source for the spectrophotometry of fluorescent specimens is today the most outstanding problem in applied colorimetry. The CIE has published a method of evaluating the closeness of any source to illuminant D65 (CIE 1981) but this requires knowledge of the SPD of the source, information which is rarely available to the spectrophotometer user in industry.

A promising solution to this problem was proposed by Clarke (1980) and Clarke and McKinnon (1982). They point out that the greatest difficulty lies in matching the SPD of D65 in the ultraviolet region, particularly below 350 nm. Most critical colour judgements, however, are made under daylight coming through window glass which attenuates the SPD in this region. The resultant SPD can be easily simulated by passing the light from a tungsten-halogen lamp through a multi-component filter made from readily available glasses. Clarke and McKinnon have therefore proposed that illuminant D65 should be augmented by illuminant/source ID65 (interior daylight) as source ID65 would, of course, be entirely suitable for incorporating in a spectrophotometer. Unfortunately, however, this proposal was rejected at the 1983 CIE meeting.

Another project which will be helpful to the spectrophotometer manufacturer is the production of a set of fluorescent reflectance standards currently under development by the Commission of the European Communities.

The use of an ultraviolet cut-off filter

Most modern spectrophotometers permit the insertion of a filter in the incident light beam which cuts off all of the ultraviolet radiation. The major use of such a filter is to determine if a white textile or paper specimen contains a fluorescent brightening agent and, if so, how effective it is. It can also be used to determine spectrophotometrically if a coloured specimen is fluorescent as the insertion of such a filter will have no effect on the reflectance values of any non-fluorescent specimen but will reduce the total radiance factors in the region of fluorescent emission. The reduction is not great, however, because most of the excitation is caused by absorbed visible light as described in chapter 2.

For quantifying the efficiency of a fluorescent brightening agent it is desirable that all UV filters should cut off at the same wavelength. The CIE method for assessing D65 simulators specifies the limit of ultraviolet as 400 nm. Filters with transmission characteristics similar to Schott KV 418 which has a transmittance of 99% at 440 nm and 0.001% at 400 nm, have been widely used for this purpose and a proposal that this should be a recommendation has been submitted to the CIE.

The processing of spectrophotometric data

The output of the 'Hardy' consisted of a reflectance curve, originally from 400 to 700 nm, the 1955 model going down to 380 nm. The fundamental method of calculating *XYZ* values from such a curve, which was described in chapter 7, is known as the *weighted-ordinate* method because calculations could be simplified by weighting, i.e. by pre-multiplying each colour matching function by the SPD of the selected illuminant. This method, however, took far too long in pre-computer days for routine use and a very much quicker method – the *selected-ordinate* method – was devised. In this method, three sets of wavelength intervals were chosen, either 30 or 100, one set resulting in the product $\bar{x}(\lambda)\,S(\lambda)$ being constant the other two sets giving constant values for $\bar{y}(\lambda)\,S(\lambda)$ and $\bar{z}(\lambda)\,S(\lambda)$. One had only to sum the reflectance factors at the selected wavelengths and multiply by a constant to obtain X, Y or Z (Hardy and Pineo 1932). Hardy devised a simple manual integrator with wavelength indicators for illuminants

A, B and C and also connected his spectrophotometer to three mechanical adding machines so that the three summations were made automatically. In 1949 this method was replaced by a more sophisticated ball-and-disc integrator (Davidson and Imm 1949). In the early 1960s spectrophotometers were interfaced to conventional card or tape punches so that the reflectance values could be fed into an off-line computer which could use the weighted-ordinate method. In 1965 a spectrophotometer was interfaced to an analog computer so that the analog equivalents of reflectance values at each of seventeen wavelengths were automatically entered into the computer (Cutler 1965). In 1969, Varian Associates (USA) and Instrumental Colour Systems (UK) independently interfaced spectrophotometers to digital computers thus setting the pattern for the future (McLaren and Perry 1979), the only subsequent modification being the replacement of the bulky minicomputer by a small microprocessor. A modern spectrophotometer will display *XYZ* values or any of the derived values to be measured in the next chapter within a few seconds of pressing the MEASURE button.

The advent of the microprocessor has markedly reduced the time required for calibration. Before 1969 this would often take over fifteen minutes but once spectrophotometers had been interfaced to digital computers the time was reduced to that necessary to measure a white tile and a black tile whose absolute reflectance factors had been input into the computer's memory. All subsequent measurements could then be automatically corrected to refer to the perfect reflecting diffuser.

The advent of the microprocessor, however, has not been entirely beneficial. Some instruments display *XYZ* values with no indication as to which observer data or illuminant has been pre-programmed. This omission can easily be rectified by the manufacturer but there is another disadvantage which cannot. Once the reflectance factors have been obtained, *XYZ* values and any derivatives can then be calculated for various illuminants. This is highly desirable if the specimen is not fluorescent but if it is the displayed values are not reflectance factors but radiance factors and the *only* valid calculations are those using the SPD of the light falling on the specimen during measurement. The instrument manufacturer cannot suppress any invalid values, however, because it is not always possible to deduce that a specimen is fluorescent by inspection of its radiance factors. Naturally if any exceed 100% it must be fluorescent but many fluorescent specimens will not even reach this value.

The concept of numerical standards

Physical standards inevitably change over long periods of time even if stored under ideal conditions between periods of use. The long-term precision of modern spectrophotometers, however, is such that it has become feasible to measure a standard as soon as it has been prepared and to store the data for subsequent use instead of remeasuring the standard. Marcus found that for periods of up to 719 days, the colour difference between a measurement and the stored data was of the same order of magnitude as the impermanence of paint standards (Marcus 1978).

It is, however, essential to monitor the precision of the instrument continuously by measuring physical standards of known permanence once a week and ensuring (*a*) that there is no systematic drift in L^*C^*h values (see next chapter) and (*b*) the average colour difference between each measurement and the stored data is appreciably less than the tolerances permitted in manufacture. The ideal physical standards are the British Ceramic Research Association (BCRA) tiles.

The BCRA tiles

In 1965 the BCRA was approached by the Colour Measurement Committee of the Society of Dyers and Colourists with a view to the production of a set of physical standards of absolute permanency in several colours. One thousand sets of twelve tiles were marketed in 1968 together with an abridged version of the calibration data obtained by the National Physical Laboratory on the Master Standards and estimates of the variations of individual standards against these (Clarke and Samways 1968). Experience has shown that the tiles which are most effective for monitoring – and also for revealing differences between instruments – are the yellow, maroon and dark blue. During 1983 supplies of these tiles became exhausted and they were replaced by a similar set known as Ceramic Colour Standards: Series II.† These included two pairs of colour difference tiles, one pair exhibiting a spectrally simple difference, the other a more complex but not metameric difference. Seven of the Series I tiles were significantly thermochromic (McLaren 1978, Compton 1984) as are seven of those in Series II. A 10°C change in temperature of the red and orange tiles, for example, causes a change in colour of 1.5 CIELAB units. It is therefore essential that comparisons are made only between measurements made at approximately the same temperature.

9.2 TRANSMITTANCE SPECTROPHOTOMETERS

Most reflectance spectrophotometers can be used to measure the transmittance of transparent solids or solutions upon which all the calculations possible for surface colours (described in the next chapter) can be performed. The most important function of a solution spectrophotometer is, however, its use in analysis and for this purpose the modern abridged spectrophotometer is valueless because measurements are only made at every 20 nm, the bandpass being at least 10 nm. For analysis, a complete scan is essential and this should include as much of the ultraviolet and infrared as possible: the narrower the bandpass, the better. The uses of absorptiometry in chemical analysis are outside the scope of this book but there are two applications of transmittance measurement which should, perhaps, be briefly mentioned as they were developed using the Hardy spectrophotometer.

† Obtainable from British Ceramic Research Association, Queens Road, Penkhull, Stoke-on-Trent, England.

The determination of strength

If a single dye is involved and the path length is the same then the Beer-Lambert law reduces to the simple relationship that the logarithm of the reciprocal of the transmittance T, expressed on a 0–1 scale, is linearly related to the concentration of dye in solution. $\mathrm{Lg}(1/T)$ is termed density or absorbance and a special cam could be fitted to the Hardy which plotted density as a function of wavelength. If two solutions contain the same amounts of the same commercial dye then the strength of the batch expressed as a percentage of that of the standard is given by

$$\frac{\lg (1/T)_{\mathrm{batch}}}{\lg (1/T)_{\mathrm{standard}}} \times 100$$

the density being measured at the wavelength of maximum absorption. These values are used to indicate the relative strengths of different brands of the same dye e.g. Red B200 is twice as strong as the standard brand usually designated Red BS. The reciprocal of this value multiplied by 10 000 is often used in strength comparisons, e.g.

$$1/200 \times 10\,000 = 50$$

therefore Red B200 is stronger 50:100 than Red BS.

Dye identification

The shape of the transmittance curve of an aqueous solution of a dye depends on three factors: (*a*) the chemical identity of the dye; (*b*) the pH of the solution and (*c*) the concentration. The variable of pH can be easily controlled and as it is completely impracticable to control the concentration of pure dye it is extremely fortunate that the effects of this variable can be easily eliminated. The Beer-Lambert law states that the density of a solution is equal to the product of an absorptivity $a(\lambda)$ which is a wavelength-dependent function of the dye, the path length b and the concentration of dye c

$$\lg(1/T) - a(\lambda)bc.$$

It therefore follows that

$$\lg \lg (1/T) = \lg a(\lambda) + \lg b + \lg c$$

so that as only the absorptivity is wavelength-dependent, the shape of the $\lg \lg (1/T)$ or log-density curve is independent of the concentration and the path length and is governed solely by the absorptivity of the dye. If the log-density curve of an unknown amount of dye A at a given pH is plotted on transparent graph paper it can be superimposed on the curve of any other amount of dye A at the same pH.

If a family of curves of known dyes at two or three controlled pH values is built up then determining the log-density curves of any unknown single dye at the same pH values will permit identification.

9.3 TRISTIMULUS COLORIMETERS

Because the first spectrophotometers were very expensive, somewhat slow and required highly skilled staff for successful operation there was a ready market for a cheaper, faster, easier-to-use instrument. This need was met by the tristimulus colorimeter which was designed to respond to spectral distributions of light in the same manner as the human eye. This was achieved by adjusting the spectral response of a photodetector so that it was spectrally equivalent to the CIE Standard Observer colour matching functions. This adjustment was usually achieved with three, or sometimes four, filters.

The first commercial instrument was developed by Hunter when he was at the National Bureau of Standards (NBS) in Washington (Hunter 1940) and in 1952 he established Hunter Associates Laboratory Incorporated to manufacture and market tristimulus colorimeters; many other manufacturers subsequently entered this field. Whilst the absolute accuracy of tristimulus colorimeters left much to be desired, for comparing two specimens using one or other of the colour difference equations to be described in chapter 10 they served a very useful need, providing rapid and precise measurements at a relatively low cost. Today, however, spectrophotometers are just as fast and as easy to use and have the important advantage of being able to deal with metamerism which the tristimulus instrument could not handle as it could only duplicate a single observer-illuminant combination. It is therefore probably obsolescent as far as fully equipped colour laboratories are concerned (Billmeyer and Hemmendinger 1981) though it may find new fields where colour difference measurement for quality control is of value and the price differential attractive. As such laboratories are outside the scope of this book no detailed account of tristimulus colorimeters will be given.

10

The Quantification of Colour Differences

If two specimens have the same tristimulus values they will be a perfect match to the appropriate standard observer when viewed under the appropriate source and viewing geometry. Conversely if any tristimulus value is different they will not match and the overall difference will be at least a rough measure of the perceived colour difference between them. The overall difference is easily quantified by regarding the XYZ values as co-ordinates in a Euclidean colour space. The difference is then the distance between standard and batch in this space which is given by the application of the Pythagoras theorem to three dimensions

$$\Delta E = [(\Delta X)^2 + (\Delta Y)^2 + (\Delta Z)^2]^{0.5}$$

where Δ signifies 'difference in' and E is the initial letter of the German word 'Empfindung' meaning *sensation*. It was soon realised, however, that the above equation was not satisfactory as some colour differences which were perceptually equal in size gave ΔE values varying by 30:1. Since the mid-1930s more effort has been devoted to solving the problem of quantifying colour differences than was involved in the development of the 1931 CIE system of colour measurement which started with Maxwell. Between 1936 and 1976 over 20 colour difference equations or formulae – the terms are synonymous – were developed. Only some of the highlights will be mentioned here and for a comprehensive survey the reader is referred to McDonald (1982).

The solution of the problem required a number of pairs of specimens exhibiting colour differences which were perceptually equal: their $xy\,Y$ values were then subject to mathematical transformations either to produce a more uniform colour space, or by some other method, to obtain the same values of ΔE. The problem was attacked in two parts with equally spaced coloured lights being used to quantify differences in chromaticness and equally spaced surface colours to quantify differences in chromaticness and lightness.

Chromaticness differences

Mathematical transformations of the xy values of equiluminous lights were first used by Judd to derive a uniform chromaticity-scale (UCS) diagram by making a linear or projective transformation of the *xy* diagram (Judd 1935). Several others were produced of which that of MacAdam (1937) became the 1960 CIE-UCS diagram, the transformation equations being

$$u = 4x/(-2x + 12y + 3)$$

$$v = 6y/(-2x + 12y + 3).$$

The second set of equally spaced visual data was obtained by MacAdam (1942) using an ingeniously designed colorimeter in which, by turning a single control knob, the colour of one half of a 2° field was varied along a straight line oriented in any direction in the *xy* diagram, its luminance being held constant. The field was surrounded by a brighter area resembling source C so that the matching field had the appearance of coloured surfaces and not that of lights. The other half of the field lay on the same straight line and had the same luminance. The standard deviations of the settings at various orientations were determined and extensive auxiliary experiments showed that a just perceptible difference was about three times as large as the standard deviation. The standard deviations fell closely on ellipses surrounding each point studied and these are shown, enlarged ten times for clarity, in figure 10.1. Each of these ellipses can be represented by the equation

$$g_{11}(dx)^2 + 2g_{12}\,dx\,dy + g_{22}(dy)^2 = 1$$

where dx is the difference between the x co-ordinate of the ellipse centre and any point on the ellipse, dy is the corresponding difference in the y co-ordinates and g_{11}, g_{12} and g_{22} are metric coefficients. MacAdam (1943) published three *xy* diagrams showing contour lines of constant metric coefficient from which the MacAdam ellipse at any *xy* value could be plotted. The distance of a specimen in relation to the distance of the ellipse in the same direction then gave the chromaticness difference which could be combined with any lightness difference to give ΔE. The geometry of such a colour space is not Euclidean but Riemannian and its application only became a practical proposition when Simon and Goodwin (1957) devised charts for different regions of the *xy* diagram in which linear distances represented chromaticness differences in MacAdam units. Other charts were devised for incorporating any lightness difference, the Union Carbide Company in the USA providing sets of these charts to any interested party free of charge.

The most important colour difference equation based on MacAdam ellipses resulted from a very sophisticated theory of colour vision developed by Friele which was based on the Müller two-stage theory mentioned in chapter 4. This could be used to predict the size, shape and orientation of the MacAdam ellipses and the predictions were remarkably close (Friele 1961). Collaboration with MacAdam (1966) and Chickering (1967) resulted in two equations, FMC-1 and FMC-2, the latter containing a function to vary the size of the ellipse according to the Y value: this became the most popular equation in the USA in spite of the fact that Friele and MacAdam always preferred FMC-1 (MacAdam 1969).

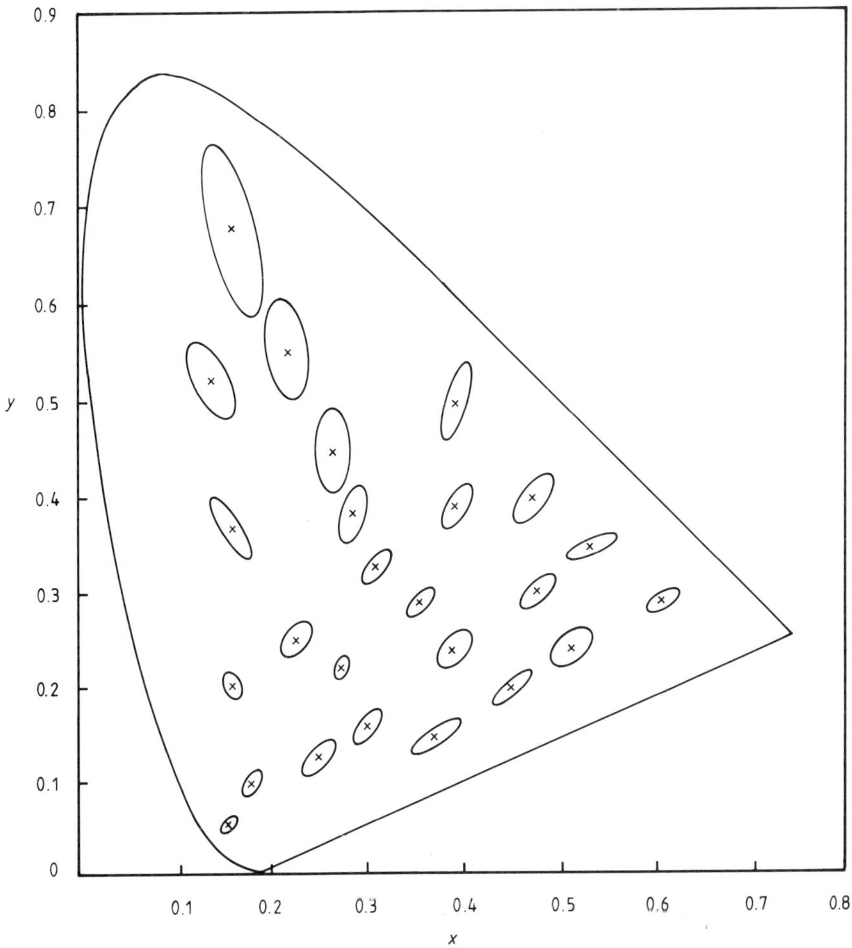

Figure 10.1 The MacAdam ellipses. Each point on an ellipse represents a colour whose difference from the colour at the ellipse centre is three times larger than just perceptible (ellipses enlarged 10 times).

Lightness differences

It was known in 1920 that measured luminance, the precursor of the CIE 1924 $V(\lambda)$ curve, could not be converted into the equally-spaced Munsell 0-10 value scale by a linear transformation. Several non-linear transformations were developed from the simplest

$$V = Y^{0.5}$$

(Priest *et al* 1920) to the most complex

$$Y = 1.2219V - 0.23111V^2 + 0.23951V^3 - 0.021009V^4 + 0.0008404V^5$$

which was devised by Judd in connection with the Munsell renotation studies mentioned in chapter 7 and is generally known as the 'Judd polynomial' (Newhall *et al* 1943).

The square-root function was used in the colour difference equation which defined the NBS unit of colour difference (Hunter 1942) which for many years was a widely accepted yardstick for colour difference assessments though the generalisation that 'differences of 1 NBS unit or less can usually be disregarded in commercial transactions' (Judd and Wyszecki 1975, p. 318) was never justified. It was also used extremely effectively in 1948 by Hunter who was able to market a tristimulus colorimeter in which electronic circuitry applied a square-root function to the *RGB* signals thus giving directly units of a uniform colour space (Hunter 1948). This anticipated by nearly thirty years the arrival of the microprocessor.

The Judd polynomial was used in the very important ANLAB equation whose origins are most conveniently considered in the next chapter. Most other investigators, however, considered that the scatter of the visual assessments of lightness differences was too great to warrant such a complicated equation especially as it was inverted, i.e. it gave Y instead of V_Y. Some investigators therefore devised cube-root approximations, the first being that of Ladd and Pinney (1955), from which various cube-root equations emerged.

The target: uniformity of practice

At the start of the 1960s there were many different equations being used with no possibility of accurate interconversion of colour difference units (McLaren 1969). In a praiseworthy attempt to achieve uniformity of practice the CIE extended the 1960 UCS diagram into three dimensions by incorporating a cube-root uniform lightness scale to give the CIE 1964 $U^*V^*W^*$ colour space and colour difference equation. This, unfortunately, failed because applications in industry showed that it was less reliable than other equations (Berger and Brockes 1966) and in 1967 the CIE recommended that three other equations be compared with CIE 1964 (Wyszecki 1968). These equations were:

(*a*) the cube-root formula of Glasser *et al* (1958) which had been modified by Reilly;

(*b*) the Munsell renotation formula (Godlove 1951);

(*c*) FMC-2.

Data for such a comparison had been obtained by Davidson and Friede (1953). About 12 000 individual judgements of the acceptability of 287 batches of carpet yarn dyed to match nineteen standards were made by eight professional shade passers. By plotting ΔE against the number of times each batch had been rated 'pass' expressed as acceptability ($\%A$) Davidson and Friede saw that the scatter of the points was lowest for MacAdam's original equation (MacAdam 1943) combined with a lightness difference component based on Munsell value. As they also published the $xy\,Y$ values for all specimens and the $\%A$ for all batches these data could be used to compare other equations such as those recommended for study by the CIE. Such a comparison was made by McLaren who was chairman of the Colour Measurement Committee of the Society of Dyers and Colourists. This body also wished to achieve uniformity of practice but felt that this would only occur if the equation recommended could be shown to be one of the most reliable. Instead of merely relying on the visual assessment of scatter, however, he determined the correlation coefficients between ΔE and $\%A$, the results being given in table 10.1.

Table 10.1 Correlation coefficients from 287 batch/standard pairs around 19 colour centres assessed on about 12 000 occasions by eight assessors.

Equation	Correlation coefficient (absolute)
†FMC-2	0.62
Simon–Goodwin	0.58
ANLAB	0.58
†Cube-root	0.58
†Munsell renotation	0.55
†CIE 1964	0.54
FMC-1	0.46

† Recommended for study by the CIE in 1967.

Although these results seemed to show that nothing better than a MacAdam ellipse equation had emerged since 1953, unfortunately the difference between 0.62 and 0.54 was not significant at the 10% level so that no equation could be recommended on the grounds of reliability (McLaren 1970a, b). At this time, however, Jaeckel and his co-workers at the Hosiery and Allied Trades Research Association in Nottingham were obtaining additional data of the same type and when these were combined with the Davidson and Friede data for 5–95% acceptability, differences of greater significance were obtained, table 10.2 (Jaeckel 1973).

A correlation coefficient of 0.67 was significantly better, at the 1% level, than 0.61 so that it was reasonable to recommend any of the cube-root equations or ANLAB on the grounds of greater reliability. The committee chose ANLAB in spite of its not being one of those recommended by the CIE mainly because it was much more widely used in the UK: it had been used, for example, to define the spacing of the Grey and Staining scales used in fastness testing mentioned in chapter 3. Another reason was the fact that there were three cube-root formulae of equal reliability and no convincing reason for believing that the one chosen by the CIE was in any way preferable. In 1971 the ANLAB equation was adopted by the textile committee of the International Organisation for Standardisation, ISO/TC38/SC1, for development as an ISO recommendation.

Table 10.2 Correlation coefficients from 775 batch/standard pairs around 31 colour centres assessed on about 40 000 occasions by 8–32 assessors.

Equation	Correlation coefficient (absolute)
Cube-root (Glasser)	0.671
Cube-root (Morton)	0.670
†Cube-root (Glasser–Reilly)	0.666
ANLAB	0.664
NBS	0.640
†FMC-2	0.610
Hunter	0.605
†CIE 1964	0.588
FMC-1	0.445

† Recommended for study by the CIE in 1967.

The ANLAB equation has been used with scaling factors of 40, 42 and 50, the first being chosen to make the size of the unit approximately equal to the NBS unit (Nickerson 1950). In consequence colour differences calculated in ANLAB (40) units were often stated to be in NBS units (e.g. Fastness Tests Co-ordinating Committee 1953) and this still occasionally occurs in the UK paint industry. It must be deprecated because conversion from one unit to another can rarely be done reliably (McLaren 1969).

At the 1973 meeting of the CIE Colorimetry Committee MacAdam proposed that the CIE should publish a recommendation based on ANLAB but using a cube-root function instead of the Judd polynomial. As such a space could not have associated with it a UCS diagram the committee also agreed to modify the 1960 *uv* diagram by making $u'=u$ and $v'=1.5v$ following Eastwood's studies (Eastwood 1973) and to extend it into three dimensions as the 1960 diagram had been extended into the CIE 1964 $U^*V^*W^*$ space, the same lightness co-ordinate being used for the cube-root and UCS spaces. The committee's initial proposals were published in 1974 (Wyszecki 1974) and at its meeting in 1975 the following were established as CIE recommendations:

(*a*) CIE 1976 $L^*u^*v^*$ colour space and colour difference formula (CIELUV) which incorporated the $u'v'$ chromaticity diagram;

(*b*) CIE 1976 $L^*a^*b^*$ colour space and colour difference formula (CIELAB). Because two equations were recommended the CIE admitted 'that the original attempt to unify current practice of colour-difference evaluations on a single colour-difference formula has again failed' (CIE 1978). This was, however, inevitable because CIELAB, which was known to be one of the most reliable equations for quantifying the differences between surface colours, could not have associated with it a UCS diagram and such a diagram was essential for those industries concerned with additive mixtures of coloured lights such as television. It was conceivable, of course, that CIELUV might have been equally or even more reliable for surface colours but by 1975 no comparisons had ever been made and the need for uniformity of practice in this area had become extremely urgent. A survey carried out by the Inter-Society Color Council of the USA in 1973 had revealed that no fewer than thirteen colour difference equations were being used, the most popular, FMC-2 being used only by a minority (31%).

The CIELAB equation is as follows

$$L^* = 116(Y/Y_n)^{1/3} - 16 \qquad \text{if} \qquad Y/Y_n > 0.008856$$

but

$$L^* = 903.3(Y/Y_n) \qquad \text{if} \qquad Y/Y_n \leqslant 0.008856$$

$$a^* = 500[f(X/X_n) - f(Y/Y_n)]$$

$$b^* = 200[f(Y/Y_n) - f(Z/Z_n)]$$

where

$$f(X/X_n) = (X/X_n)^{1/3} \qquad \text{if} \qquad X/X_n > 0.008856$$

but

$$f(X/X_n) = 7.787(X/X_n) + 16/116 \qquad \text{if} \qquad X/X_n \leqslant 0.008856$$

$$f(Y/Y_n) = (Y/Y_n)^{1/3} \qquad \text{if} \qquad Y/Y_n > 0.008856$$

but

$$f(Y/Y_n) = 7.787(Y/Y_n) + 16/116 \qquad \text{if} \qquad Y/Y_n \leqslant 0.008856$$

$$f(Z/Z_n) = (Z/Z_n)^{1/3} \qquad \text{if} \qquad Z/Z_n > 0.008856$$

but

$$f(Z/Z_n) = 7.787(Z/Z_n) + 16/116 \qquad \text{if} \qquad Z/Z_n \leqslant 0.008856.$$

X_n, Y_n and Z_n are the tristimulus values of the perfect reflecting diffuser (which are the same as those of the illuminant), the appropriate values being

Illuminant D65/10°	$X_n = 94.8$	$Y_n = 100.0$	$Z_n = 107.4$
Illuminant D65/2°	$X_n = 95.0$	$Y_n = 100.0$	$Z_n = 108.8$
Illuminant C/10°	$X_n = 97.3$	$Y_n = 100.0$	$Z_n = 116.1$
Illuminant C/2°	$X_n = 98.1$	$Y_n = 100.0$	$Z_n = 118.2$.

Colour differences in CIELAB units are then given by

$$\Delta E = [(\Delta L^*)^2 + (\Delta a^*)^2 + (\Delta b^*)^2]^{0.5}.$$

The reason why tristimulus ratios, e.g. X/X_n, are used is in connection with the quantification of colour appearance and will be discussed in the next chapter as will the conversion of a^*b^* into values quantifying the variables of perceived colour – chroma, C^* and hue, expressed as an angular measure, h.

The use of a linear function for ratios equal to or less than 0.008856 is to overcome the defect of all cube-root functions: L^* goes negative at low Y values and significantly distorts ΔE values. This modification (Pauli 1976), however, may cause hue angles to be as much as 35° in error (McLaren 1980).

When applied to the data used for table 10.2, the CIELAB equation gave a correlation coefficient of 0.672 and this proven reliability undoubtedly played a large part in its widespread adoption by the industries concerned with surface colours. The textile industry, for example, decided to develop CIELAB rather than ANLAB as an ISO recommendation in 1974: it is now part of ISO Standard 105. Fortunately (for the protagonists of CIELAB) CIELUV has not been found to be more reliable. Four studies have shown no significant differences between CIELAB and CIELUV (Robertson 1977, Kuehni 1977, Pointer 1981, McLaren 1981) whilst one showed that CIELAB was significantly more reliable (Lozano 1977).

The quantification of metamerism

We saw in chapter 6 that two specimens having different reflectance curves could produce identical excitations of the R, G and B cones and would therefore appear to be a perfect match. Such specimens are termed 'metameric'. The degree of metamerism is normally assessed subjectively by inspection under different sources but in 1971 the CIE published an objective method, the *Special Metamerism Index: Change in Illuminant* (CIE 1971). If the tristimulus values of two specimens under D65 are identical, i.e. $X_1 = X_2$, $Y_1 = Y_2$ and $Z_1 = Z_2$, the index is simply the colour difference under the specified test illuminant. If they are not identical then 'a suitable account should be taken of this failure'. Physical specimens which are a 'good commercial match' will rarely have identical values and the most widely used method of allowing for this is to multiply the tristimulus values of the second specimen under any illuminant by X_1/X_2, Y_1/Y_2 and Z_1/Z_2 respectively under D65 (Brockes 1970).

Lamps specially suited for detecting metamerism

In 1974 Thornton calculated the wavelengths of the crossover points of metamers and found them to be centred on 448, 538 and 612 nm. He then claimed that a three-band lamp whose emission bands were at these wavelengths would have the desirable property of preserving matches made in daylight (Thornton 1974). This reasoning was later shown to be false as the crossover points of nearly a thousand metamers generated from three-dye combinations of twelve dyes were distributed over the whole spectrum and only 1.5% had points in the region of these wavelengths (Ohta and Wyszecki 1977, McLaren 1978b). Such lamps are in fact particularly effective for detecting metamerism as many

UK dyers found to their cost when the Philips three-band lamp – Colour 84 – was installed in a large department store (Colour Measurement Committee 1976). Thornton also reasoned that a three-band lamp whose emission peaks fell *between* these wavelengths would be an effective metamer-detector and such a lamp is marketed by Westinghouse as 'Color-Match'. It is extremely unlikely that any metamerism between a pair of specimens which are an acceptable match under daylight, Colour 84 and Color-Match will ever cause any problems. The SPD of these two lamps could therefore be used to calculate CIE metamerism indexes; specimens specially prepared to illustrate extreme metamerism will have maximum indexes up to 15 and matches with a maximum index of less than 2 have never been found to cause customer complaints on the grounds of metamerism.

The alternative definitions of metamerism

Specimens A and B on the frontispiece to the second edition of *Principles of Color Technology* (Billmeyer and Saltzmann 1981) will be considered by colourists to be an excellent example of the metamerism occurring in practice as distinct from those examples deliberately prepared to illustrate extreme metamerism. This is because in daylight they are a very close match to most observers but B is much darker than A under tungsten light and very much lighter under the Philips three-band lamp, Colour 84.

A colour physicist, on the other hand, would probably deny that they were metamers according to the CIE definition of the term which is 'spectrally different radiations (colour stimuli) that result in the same psychophysical colour', the latter being defined as 'characteristics of a colour stimulus consisting of three values such as three tristimulus values'. It has been shown mathematically that if identical tristimulus values are given by non-identical reflectance curves they must cross in at least three points (Stiles and Wyszecki 1968): the curves of A and B, however (figure 10.2) cross in only one point. An index which quantified metamerism according to the CIE definition of the term should therefore give A and B zero rating. Applying the multiplicative correction just described to allow for the fact that $X_1 \neq X_2$, $Y_1 \neq Y_2$ and $Z_1 \neq Z_2$ (D65), the CIE Special Metamerism Index rates A and B as 2.3 (illuminant A) and 6.0 (Colour 84). The CIE index therefore quantifies metamerism as understood by the colourist rather than as defined by the CIE.

In computer match prediction, the subject of chapter 12, any prediction which does not yield a reflectance curve identical with that of the target shade will yield a metameric match as defined by the CIE, i.e. with three or more crossover points. Matches with fewer than three crossover points, such as A and B, are quite common when the traditional (visual) method of colour matching is used and the colorants used to colour the batch are not identical with those present in the standard. As far as industry and commerce are concerned there is no need to differentiate between the two types of metamerism but the question

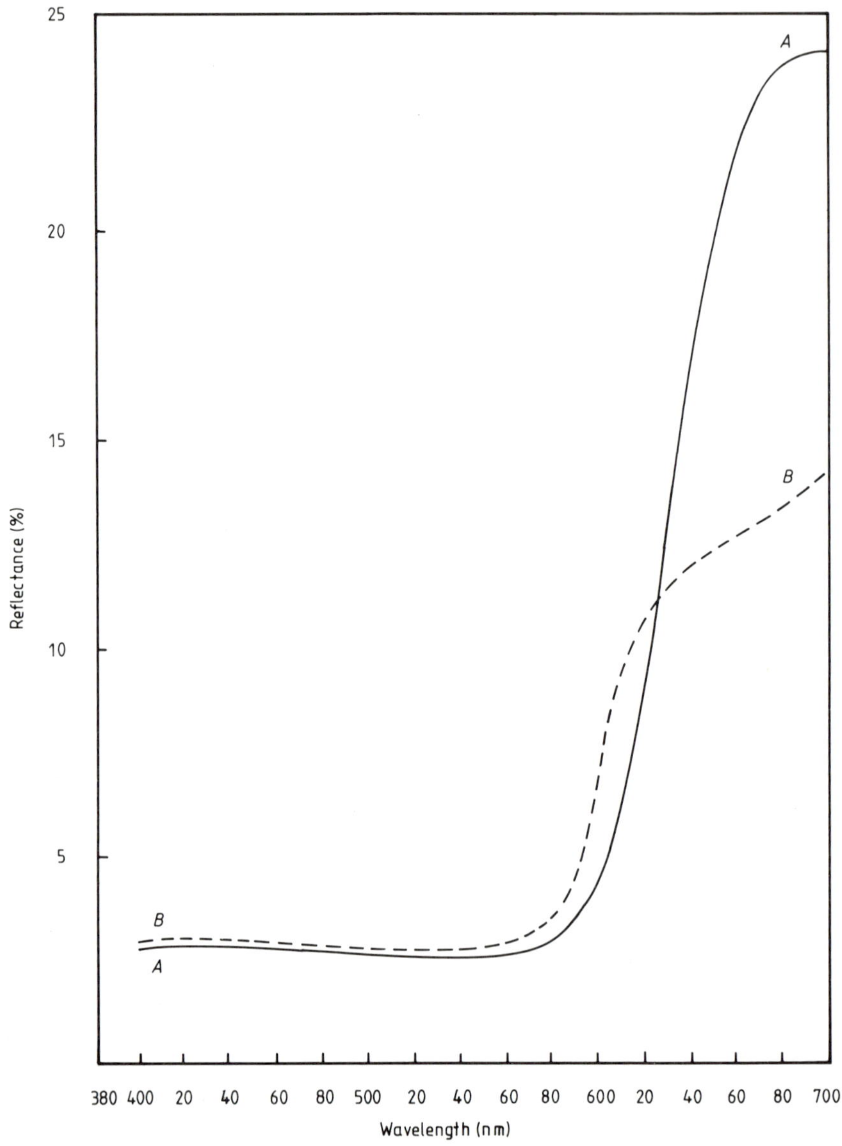

Figure 10.2 Reflectance curves of non-psychophysical metamers.

then arises as to whether another term should be used to cover both types. There are, however, some reasons against such a course:

(*a*) the term 'metamerism' was coined by Ostwald in 1919 to describe the visual phenomenon, not the psychophysical (Ostwald 1919);

(*b*) definitions by authorities other than the CIE, e.g. the British Standards

Institution (BS 4727) and the Society of Dyers and Colourists are based on the visual phenomenon;

(*c*) the term is so firmly entrenched that if any other was considered preferable, the chances of it being widely adopted are negligible.

On balance, therefore, it seems that the term metameric should continue to be used for both phenomena: when it is necessary to differentiate between them the terms *psychophysical metamers* and *non-psychophysical metamers* could be used.

Apart from the major difference between psychophysical and non-psychophysical metamers, however, there are many other differences of opinion, especially among colourists and these have been highlighted (Rodrigues and Besnoy 1980). This was followed by the American Inter-Society Color Council widely circulating a questionnaire in which 44 pointed questions on the subject were asked. Detailed comments on the problem have been made by Robertson (1983), Kuehni (1983) and Billmeyer (1983) who, between them, have proposed the following additional terms – parametamerism; paramerism; metachromic; isochromic; parachromic and orthochromic – so the responsible ISCC committee will have an extremely difficult task trying to resolve the situation. One thing that is certain, however, is that a shade which is green in daylight but brown under tungsten light must not be described as metameric as this term should only be applied to a pair of stimuli. Such a single specimen has a very low degree of *colour constancy* and this will be discussed in the next chapter.

The assessment of the absolute reliability of colour difference formulae

The use of the correlation coefficient between $\%A$ and ΔE was criticised on the grounds that the relationship between them was far from being linear (Thurner and Walther 1969). Coates *et al* (1972) therefore converted $\%A$ to a linear scale and then applied three measures of fit. McLaren (1973a) fitted polynomial and sin/log curves to $\%A/\Delta E$ data – and used 'goodness of fit' as the criterion of relative reliability. Neither of these two, more sophisticated methods significantly changed the ranking order of the formulae to which they were applied. None of these methods, however, indicates if the most reliable formula is capable of being used for single-number shade passing, in other words where a single pass/fail value can be used whose size depends solely on the closeness of match required to satisfy the customer. This criterion can best be met by comparing instrumental with visual wrong decisions.

The 'wrong decisions' criterion

When a number of assessors, ostensibly working to the same tolerance, have made replicate pass/fail decisions the correct decision must be that of the majority of assessments and all the opposite decisions must be wrong. Wrong decisions are by no means caused solely by the inability of an assessor to make the same decision on every occasion: individual assessors may have a personal tolerance which is different from the corporate tolerance. These different causes were clearly revealed in a study carried out by McDonald in which eight assessors made five separate pass/fail decisions on each of 640 batches dyed against 54 standards. A pair of grey-dyed yarns indicating the company's tolerance was in the field of view at all times. He found that whilst the range of wrong decisions as judged against the corporate decision was from 10.2 to 16.4%, it was only from 4.0 to 10.9% when judged against the assessor's personal tolerance. The most lenient assessor had a pass rate of 41%, i.e. the batches passed by the observer were passed on 41% of the total assessments and the most severe had a pass rate of 66% (McDonald 1980a).

Instrumental wrong decisions are assessed by setting a low pass/fail value of ΔE and comparing the instrumental decision with the majority visual decision. This value is then increased stepwise and the procedure repeated until the minimum number of wrong decisions has been obtained.

Visual wrong decisions occurring in the Davidson and Friede study were 17.1% and using the optimum pass/fail value for the ANLAB equation (1.3) the instrumental wrong decisions were greater, 20.2% (McLaren 1970a, b). This suggested that it would be prudent to assume that single-number shade passing using the most reliable equation developed by 1969 would not be viable. The CIELAB equation was later found to give virtually the same number of wrong decisions. Instrumental shade passing had, of course, been successfully used for many years by setting colour-specific pass/fail values but the single-number objective was still a possibility: if the colour-specific pass/fail values varied systematically according to the position of standard in ANLAB, space methods could be devised for modifying ΔE values accordingly.

The development of optimised equations, 1971–83

The first attempt to do this was made by McLaren in 1971. He took each of the nineteen colour sets in the Davidson and Friede data, fitted the best curve to $\%A/\Delta E$ and predicted the ΔE value corresponding to $50\%A$, i.e. the pass/fail boundary. These values varied from 0.57 to 2.27 and were found to vary systematically with the lightness, chroma and hue angle of the standard. The technique of multiple linear regression enabled a factor to be devised which modified ΔE accordingly and this reduced the number of wrong decisions to 15.7%, i.e. less than visual which proved to be a promising start.

The results of this approach were presented at the Helmholtz Memorial Symposium on Color Metrics held in Holland in 1971 (McLaren 1971b) and every subsequent even-dated year saw the publication of another equation optimised against one or more data sets. Each one of these gave fewer wrong decisions than visual when applied to the Davidson and Friede data even if that data was not used for optimisation (Kuehni 1972, McDonald 1974, McLaren 1976a, Friede 1978, McDonald 1980b). Three of these will be referred to specifically: $\Delta E(Mc)^2$, JPC79 and M and S/ICS.

$\Delta E(Mc)^2$

Many dyehouses using colour-specific pass/fail values of ANLAB ΔE found that it was often necessary to set a separate, lower value for Δhue. McLaren therefore determined the optimum values for l, c and h in the ANLAB equation

$$\Delta E = [(l\Delta L)^2 + (c\Delta C)^2 + (h\Delta H)^2]^{0.5} / (1 + 0.02C)$$

where the divisor was the positional modifying factor devised by McDonald (1974). Four sets of acceptability data were used including that of Davidson and Friede and the optimum values were

$$l = c = 1 \qquad h = 2.$$

This gave 12.5% wrong decisions when applied to the Davidson and Friede data (McLaren 1976a).

JPC79

At the end of 1976, McDonald prepared 640 matches around 55 colour centres, the match distribution being similar to that of 567 routine dyeings submitted for assessment during a one-month period. These were assessed five times by eight professional shade passers and some of the results have already been mentioned. From the colorimetric data of the batch/standard pairs, McDonald fitted ellipsoids to define pass/fail boundaries whose axes were oriented along lightness, chroma and hue planes and whose dimensions were optimised to minimise instrumental wrong decisions. The length of the chroma semiaxis increased with increasing chroma, and that of the lightness semiaxis with increasing lightness. McDonald also found that the optimum length of the hue semiaxis also increased with increasing chroma but that this also depended on the hue angle of standard. At the same chroma the optimum hue semiaxis dimension varied by as much as 50% according to the hue angle though when colours near the achromatic axis were involved, better agreement with visual data occurred by making the hue axis equal in length to the chroma axis.

Equations were then devised for calculating ΔE in relation to the size of the ellipsoid at the position of standard in ANLAB space which gave only 11.5% instrumental wrong decisions against the Davidson and Friede data (McDonald 1980c).

During this period the dyehouse manager at J and P Coats had been recording his pass/fail decisions on routine dyeings, and the batch/standard pairs were measured. After a two-year period data on 8454 matchings around 599 colour centres became available for optimisation studies. The severity and precision of this observer were known to be within the range of the eight used in the earlier study. Because his assessments covered a very much wider range of colours this data permitted the effect on the optimum size of the hue semiaxis to be determined with much greater precision and optimum ellipsoids were therefore fitted to this data. Equations were then derived which enabled the lengths of the lightness, chroma and hue semiaxes to be calculated from the L, C and h^0 values of standard. A colour difference was then calculated by determining how far a batch was from the standard in relation to the distance of the ellipsoid surface in the same direction.

This formula gave 11.5% wrong decisions when applied to the Davidson and Friede data and 9.7% when applied to the J and P Coats data based on 617 batches (23 or the original 640 were not available for measurement). This latter figure is not only less than the average 13.2% wrong decisions: it is less than the wrong decisions made by the most reliable of the eight observers (10.2%) and only just outside the range of wrong decisions which occurred when the visual data were treated as though they came from a panel of eight observers repeating its assessments five times, namely 5.4–9.5%.

This formula had, however, been computed for ANLAB(50) co-ordinates and unit values represented the particular J and P Coats tolerance used (equal to a lightness difference of 1.7 CIELAB units). In order to make it more universally applicable, two modifications were necessary.

The first modification was to use a scaling factor of 43.909 for the ANLAB equation as this gives virtually the same values as CIELAB: in particular when $Y = 100.00$, $L = 100.00$, when $Y = 0$, $L = 0$. The second modification was required to scale the size of ΔE_{JPC79} units: several alternatives were considered and the eventual choice was to make the average optimum pass/fail value in the Davidson and Friede data 1.0 units (McDonald 1980b).

The final JPC79 equation was as follows, the lengths of the ellipsoid semi-axes being given by

$$SA_L = 0.08195L/(1 + 0.01765L)$$

$$SA_C = 0.0638C/(1 + 0.0131C) + 0.638$$

$$SA_H = t \times SA_C$$

where $t = 1$ if either C_{standard} or C_{batch} is less than 0.638. Otherwise

$$t = 0.36 + |0.4 \cos(h^0 + 35)|$$

unless h^0 is between 164 and 345, when

$$t = 0.56 + |0.2 \cos(h^0 + 168)|$$

$$\Delta E_{JPC79} = [(\Delta L/SA_L)^2 + (\Delta C/SA_C)^2 + (\Delta H/SA_H)^2]^{0.5}.$$

This formula, unfortunately, had a discontinuity as one approached the achromatic axis and the length of the lightness semi-axis when the L^* value of standard fell below 16 had been obtained by extrapolation – to zero – in the absence of any visual data. The former meant that ΔH values could change by a factor of 4 on repeat measurements made with a high-precision spectrophotometer. The latter defect meant that ΔE values for very dark specimens were often much larger than the acceptability of the match would require. Subsequent studies eliminated both of these defects resulting in a better formula (Clarke *et al* 1984) but before this is described it is necessary to consider a long-standing controversy – is the *acceptability* of a match dependent only on the perceived size, or *perceptibility*, of the colour difference between standard and batch or are some equally perceptible differences more readily acceptable than others?

The perceptibility/acceptability controversy

It is widely accepted that if the size of the perceived colour difference between standard and batch A is the same as that between standard and batch B, A and B will not necessarily be given the same acceptability rating if the difference is near the assessor's pass/fail boundary. In particular, if the colour difference between standard and A is mainly one of hue and that between standard and B is mainly one of lightness and/or chroma, B is more likely to be accepted. Varying the weightings given to the lightness, chroma and hue difference components of the ANLAB formula applied to three sets of acceptability and one set of perceptibility data confirmed this (McLaren 1976b).

When Clarke *et al* had eliminated the two defects in the JPC79 formula they also varied the weightings and found that for maximum agreement with four sets of perceptibility data the contribution of the lightness difference component should be doubled. As assessments of perceptibility are invariable, unlike those of acceptability which probably vary from industry to industry, the Colour Measurement Committee of the Society of Dyers and Colourists, who had sponsored this work, decided to publish a formula containing optional weighting factors for lightness, chroma and hue differences so that for quantifying perceptibility these factors would be unity. This would have meant that for quantifying acceptability the factors would often be less than unity and to avoid the necessity for a decimal point the reciprocals of the weighting factors were used, designated l for lightness differences and c for chroma differences. One factor has to be fixed at unity otherwise the size of ΔE could vary with identical weightings: hue differences were thought to be least susceptible to variations between industries. l and c are therefore *relative tolerances* for lightness and chroma differences respectively.

The final formula was called CMC$(l:c)$ after the initials of the sponsoring committee and is as follows:

Input CIELAB values L^*C^* and h for standard CIELAB colour difference components $\Delta L^*\Delta C^*\Delta H^*$

Calculation of ellipsoid semi-axes $S_L S_C$ and S_H

$$S_L = \frac{0.040975L^*}{1 + 0.01765L^*} \qquad \text{unless } L^* < 16 \text{ when } S_L = 0.511$$

$$S_C = \frac{0.0638C^*}{1 + 0.0131C^*} + 0.638$$

$$S_H = (fT + 1 - f)\,S_C$$

where

$$f = \left(\frac{(C^*)^4}{(C^*)^4 + 1900} \right)^{0.5}$$

and

$$T = 0.36 + | 0.4 \cos(h + 35) |$$

unless h is between 164° and 345° when

$$T = 0.56 + | 0.2 \cos(h + 168) |.$$

Calculation of colour difference

$$\Delta E = \left[\left(\frac{\Delta L^*}{S_L} \right)^2 + \left(\frac{\Delta C^*}{S_C} \right)^2 + \left(\frac{\Delta H^*}{S_H} \right)^2 \right]^{0.5}$$

$$= [(\Delta L)^2 + (\Delta C)^2 + (\Delta H)^2]^{0.5}.$$

When CMC(1 : 1) was compared with CIELAB against the four sets of perceptibility data it gave a significantly higher coefficient of variation for two of them and a significantly lower coefficient for the other two. The most likely reason is the size of the average colour difference as shown in table 10.3.

Table 10.3 Coefficient of variation and average size of colour difference for four sets of perceptibility data.

Set	Coefficient of variation (%) CMC(1 : 1)	CIELAB	Average size of difference
1	32	51	2.3
2	39	43	5.2
3	26	19	14.3
4	44	26	20.9

This shows that if CIELAB space is non-uniform for large colour differences it cannot be made more uniform by the transformations which significantly improve its uniformity for small differences. This could mean that CIELAB space is reasonably uniform and this would be very important because the colour space associated with the CMC($l : c$) formula is not Euclidean and therefore cannot be represented by a real model. CIELAB space is undoubtedly more uniform for large differences than Munsell space whose spacing is uniform within Value, Chroma and Hue planes but not between planes as described in chapter 7 (McLaren 1983b).

CMC(1 : 1) is therefore being investigated by the CIE as a possible replacement for CIELAB for quantifying the perceptibility of small colour differences and also by ISO for assessing the effect on the specimen in fastness testing which will be described later.

In spite of this additional confirmation that there is a fundamental difference between acceptability and perceptibility assessments this is still being denied (Kuehni 1984).

The M&S formulae

In 1975, Taylor of Marks and Spencer (a large merchandising chain) discussed the 'optimised equation' approach to the problem of single-number shade passing with McDonald and his colleagues and began an independent attack on the problem (Taylor 1977). In collaboration with Smart of Instrumental Colour Systems (Newbury, England) a colour difference formula was developed which, unlike JPC79, could be 'fine-tuned' to eliminate any weaknesses which extensive industrial usage revealed. This resulted in several major modifications which were identified by two digits indicating the year. Publication would have been premature as long as fine-tuning continued but it is expected that the latest, known as M&S 83A, will be the last. Absence of publication, however, has not prevented M&S 83A becoming the most widely used optimised formula throughout the world, mainly, but not exclusively, in the textile industry. It resulted in the two companies jointly receiving a Queen's Award for technological achievement in 1984.

The Datacolor formula

The most recent optimised equation was devised by Datacolor (Rohner, private communication) using CIELAB colour differences L^*, C^* and H^* for between 60 and 120 acceptable matches of dyed yarns against standards of 273 different colours. Lightness, chroma and hue tolerances were then calculated as functions of the L^*, C^* and h co-ordinates of standard.

The reliability of the optimised equations

By 1980 the pioneering studies of Davidson and Friede had been augmented by many others giving a grand total of over 100 000 assessments made by over 100 professional shade passers involving more than 10 000 batches assessed against standards of over 800 different colours. Details of these studies are given in table 10.4.

The visual wrong decisions in these studies varied from 13.1 to 24.2% so that the best method of indicating the performance of any equation is to calculate % instrumental minus % visual wrong decisions. The more negative the difference, the greater the reliability of the equation. The differences resulting from the use of four equations are shown diagrammatically in figure 10.3.

Table 10.4 Sets of acceptability and perceptibility data.

Set	No. of standards	No. of batches	No. of assessors	Reference
D and F	19	286	8	Davidson and Friede (1953)
HATRA	12	589	24	Jaeckel (1973)
K/M	13	289	10/16	Kuehni (1971a, b)
JPC 169	169	169	8	–
JPC 617	55	617	8	McDonald (1980a, b, c)
JPC 8454	599	8454	1	McDonald (1980c)
MMB	19	518	20	Morley *et al* (1975)
ISCC	6	179	26	Kuehni and Marcus (1979)

The CMC(2 : 1) and M&S 83A formulae have been applied to five of these data sets but do not show any significant differences. This was to be expected because the undoubted improvements apply to only a very small proportion of the colours used in each set.

These results confirm that CIELAB is not reliable enough for single-number shade passing but that the use of any one of the three optimised equations will result in fewer wrong decisions than are made by the average professional colourist making a single visual assessment. Of the three, JPC79 and hence CMC(2 : 1) appears to be the most reliable but the simple wrong decisions criterion may not be adequate for differentiating between equations which are significantly better than CIELAB. This is because it does not take into account the seriousness of the wrong decision.

One method of doing this is to calculate for each instrumental wrong decision what change in ΔE is required to make the decision the correct one. This more sophisticated approach has so far been applied to only one set of data (unpublished). Five colourists assessed 2561 batches dyed against standards of 285 different colours. The optimum pass/fail value was determined for CMC(2 : 1) and M & S 83A and for each instrumental wrong decision its seriousness was assessed by determining the change in ΔE required to give the right decision. As the optimum pass/fail value was different for each equation this change was expressed as a percentage of this value. In the case of instrumental fail/visual pass wrong decisions the distribution pattern for both formulae were similar but in the case of the more serious instrumental pass/visual fail wrong decisions the patterns were quite different as shown in figure 10.4. This shows that for this, the more serious type of instrumental wrong decision, the use of CMC(2 : 1) results in significantly more wrong decisions than M&S 83A (58 as against 30) and the wrong decisions are more serious.

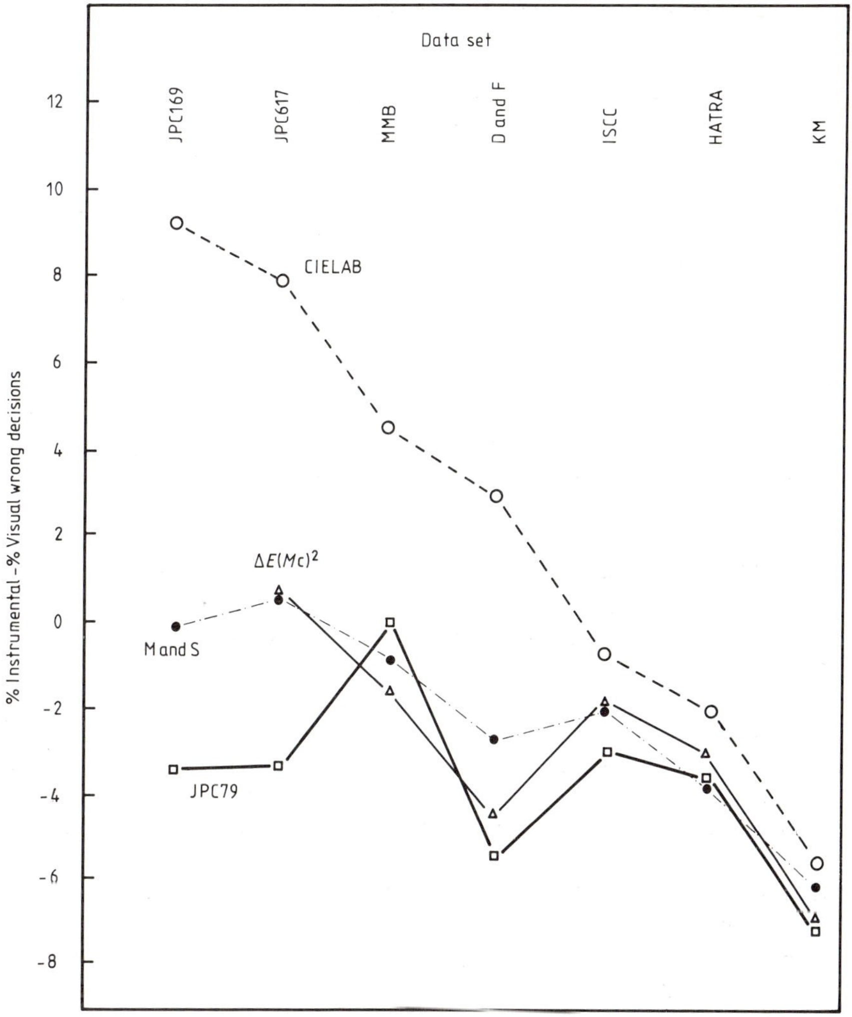

Figure 10.3 Reliability of various equations in quantifying perceived colour differences.

An even better method would take into account the fact that if a batch has an acceptability rating of 45–55% it is obviously of borderline acceptability and any wrong decision is therefore of no consequence. On the contrary, an instrumental fail for a batch with an acceptability rating of 80–100% is serious as it will lead to expensive and possibly unsuccessful attempts at shading. An instrumental pass for a batch with an acceptability rating of 0–20% is much more serious as it will result in goods being returned for correction and the reputation of the dyehouse will suffer.

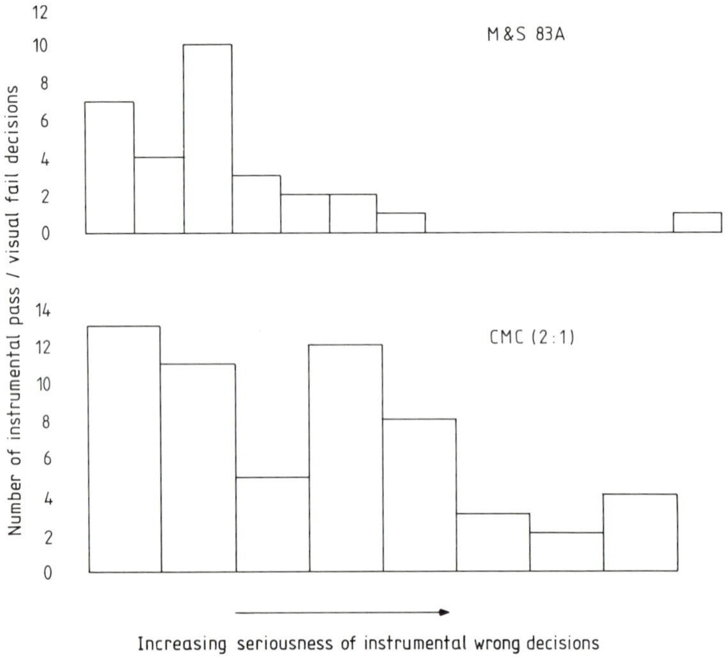

Figure 10.4 Numbers of instrumental wrong decisions of varying seriousness with CMC(2 : 1) and M&S 83A.

Such an analysis will have to be applied to all the available data sets before the most reliable optimised equation can be identified.

Shade sorting

The most efficient method of garment manufacture is to form a large number of layers of fabric on a cutting table and to cut with a band saw the component panels minimising waste. Each piece used to form the cutting 'lay' has, of course, been previously accepted as a 'good commercial match' against the appropriate standard but a colour difference which is acceptable between garments of the same nominal colour on a display rack may be entirely unacceptable between adjacent panels in a garment. The simplest method of preventing this happening is to place a layer of tissue paper between each piece and to use only panels within the same paper sandwich for the same garment. The cost of the paper is, however, by no means negligible and small panels often escape from the sandwich and have to be treated as waste.

The first method of eliminating these defects was to have a specimen from each piece assessed visually and grouped into cutting lots so that no pair within a lot would exhibit a colour difference which was unacceptable between adjacent panels. Each cutting lot could then be cut separately thus eliminating the need for paper separation or this could be minimised by using paper between cutting lots. Visual sorting is, however, the most difficult of all the visual assessments of colour and as it is the most boring it is becoming increasingly difficult to replace visual sorters on retirement. It is also slow and expensive and results in too many cutting lots as the reliable sorter invariably 'plays safe'. It is, however, a task which can be done instrumentally as it is only an extension of instrumental shade passing and the first method was devised in the 1960s to rationalise batch-to-batch variations in spun-dyed Dynel. The method was based on measurements made with a Hardy recording spectrophotometer plus mechanical tristimulus integrator. The resultant *XYZ* values were converted manually and plotted on a transformed MacAdam *xy* diagram to cover differences in chromaticness, differences in lightness being covered by a weighted Munsell Value scale (Simon 1961). Subsequently one or two fully instrumental methods based on analog devices were marketed but none were particularly successful and it was only with the availability of digital computers in the 1970s that instrumental shade sorting became an important part of quality control.

The first shade sorting system using a digital computer was installed at the Standish Company's dyehouse in Wigan in 1971 (McLaren 1975). This company dyes cotton/polyester pieces continuously and variations between pieces meant that at the end of the dyeing operation there were often unacceptable colour differences between pieces necessitating sorting visually. Specimens from both ends of each piece were measured on a Harrison–Shirley tristimulus colorimeter, the operator reading out the *XYZ* values which another operator punched on to tape which was then read into a remote computer over the telephone network. The major requirement was to obtain a cutting lot containing the greatest number of pieces so that once every piece had been located in ANLAB space a sphere whose diameter represented the maximum permissible colour difference between panels in a garment was inserted mathematically into the cloud of points and moved systematically until it contained the greatest number. If additional cutting lots were required, the points representing the first lot were removed from the cloud and the operation repeated. The method was successful and the colorimeter was then interfaced to a minicomputer so that only one operator was required (Cooper 1975).

The next development occurred in 1975 when Hunter Associates Laboratory Inc. interfaced a programmable calculator to their tristimulus colorimeters which was programmed to carry out shade sorting in Hunter *Lab* space. This was divided into hollow rectangular prisms whose faces were oriented along *L*, *a* and *b* planes, the *L* dimension being much greater than the other two. The centroid block was designated 555, the position of other blocks being designated by incremental increases or decreases in *L*, *a* and *b*: the block designated 647 was thus centred one unit lighter, one unit less in *a* and two units greater in *b*, this being the method originally introduced by Simon.

In 1976 Burlington Industries modified their shade sorting system which was based on rectangular prisms in CIELAB space whose faces were oriented along L^*, a^* and b^* planes in favour of one in which the orientation was along L^*, C^* and *h* planes (Acree 1983). The actual program was sent to several organisations and became widely known as 'Hue Angle Sort'. Orientation along planes representing the three variables of perceived colour is much more effective than orientation along only one (lightness) and the best overall ratio between the three dimensions was found to be 4:2:1, that of the common American brick (Simon 1983). An equivalent orientation had also been used in Simon's original (1961) method.

It is obvious from McDonald's mapping of the equally acceptable surfaces in *Lab* space that the size and aspect ratio of the block used must vary if sorting is to be reliable and is not to result in an unnecessarily large number of cutting lots when many different colours are involved. It is equally obvious that this complication can be avoided by basing the sorting method on the values of ΔL, ΔC and ΔH calculated using an optimised equation. At the time of writing (July 1985) such sorting programs based on M&S 83A only are currently in use and this has made for even greater efficiency in sorting as well as greater simplicity. The reason is now explained.

The maximum diameter of the shade sorting volume must represent the maximum permissible colour difference between panels in a garment and the most efficient, that having the greatest volume, is a sphere. Unfortunately spheres cannot be used because they do not close pack and therefore 26% of batches will fall in the interstices and will not be sorted. The conventional brick-shaped block becomes a cube in a uniform colour space but although this does close pack it has only 36.8% of the volume of a sphere of the same diameter. This block therefore results in very inefficient sorting. There are, however, several regular solids which close pack but have significantly greater volumes for a given maximum diameter. The first of these solids to be used in a commercially available shade sorting programme was the 12-faced rhombic dodecahedron, figure 10.5(*a*) whose volume is 47.7% that of the sphere (Smart 1977) but this has now been replaced by the 14-faced truncated octahedron, figure 10.5(*b*), the solid, capable of close packing which has the maximum volume for a given maximum diameter. Its volume is 68.3% of that of the sphere. If the batch distribu-

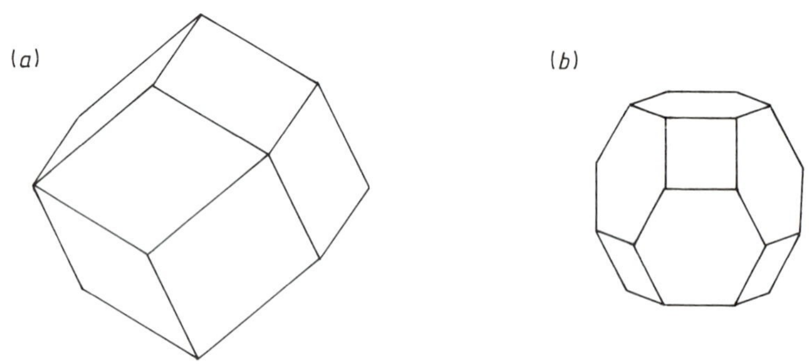

Figure 10.5 Solids which close-pack and have the greatest volume for a given diameter. (*a*) Rhombic dodecahedron, (*b*) truncated octahedron.

tion is uniformly distributed in colour space this will theoretically result in 46% fewer cutting lots than will be given by cubic sorting, with no loss in reliability.

Colour difference measurement in fastness testing

The final stages in a fastness test, described in chapter 3, involve the visual estimation of colour differences, that between the original and the tested specimen and that between an adjacent fabric present during the test and a specimen of the untreated fabric. The magnitudes of the differences are evaluated against two scales illustrating lightness differences. The scale for assessing the effect on the specimen, commonly called the 'grey scale', consists of five (or nine) pairs, one member of each pair being a medium grey.

The second member illustrating grade 5 is identical, those illustrating grades 4–5 to grade 1 being increasingly lighter. The scale for assessing the degree of staining, commonly called the 'staining scale', also consists of five (or nine) pairs, the reference member being white and the second members being increasingly darker. The spacings were originally defined in ANLAB (40) units but the CIELAB values are given in table 10.5, the bracketed grades being optional.

It is obvious that the visual assessment stage could be replaced by colour difference measurement but direct conversions, e.g. if a tested specimen has changed by 3.4 CIELAB units its grey scale rating is 3, are seriously in error because of the non-uniformity of CIELAB space. An optimised equation has

Table 10.5 Spacing of scales used in fastness testing.

Scale	L^* value of reference	Grade	ΔE (CIELAB units)
Grey scale	39.6–42.8	5	0
		(4–5)	0.8
		4	1.7
		(3–4)	2.5
		3	3.4
		(2–3)	4.8
		2	6.8
		(1–2)	9.6
		1	13.6
Staining scale	$\nless 93.9$	5	0
		(4–5)	2.3
		4	4.5
		(3–4)	6.8
		3	9.0
		(2–3)	12.8
		2	18.1
		(1–2)	25.6
		1	36.2

been developed for assessing the degree of staining and another for assessing the effect on the specimen.

Instrumental assessment of staining

The colour difference in CIELAB units between a specimen of the untreated adjacent fabric and that involved in the fastness test is converted into the staining scale rating (SSR) by means of the equation

$$\text{SSR} = 7.05 - 1.43 \ln (4.4 + \Delta E_{\text{CIELAB}})$$

and the conversions given in table 10.6 (Fastness Tests Co-ordinating Committee 1983).

Instrumental assessment of the effect on the specimen

Visual assessments in fastness testing are assessments of perceptibility not acceptability and the CMC(1:1) formula can be used to predict grey scale ratings using the conversions given in table 10.7.

Table 10.6 Conversion of calculated SSR values to reported ratings.

Calculated SSR	Reported SSR
>4.87	5
4.86–4.25	4–5
4.24–3.75	4
3.74–3.25	3–4
3.24–2.75	3
2.74–2.25	2–3
2.24–1.75	2
1.74–1.25	1–2
<1.25	1

Table 10.7 Conversion of calculated colour differences to reported grey scale ratings.

Colour difference CMC(1:1) units	Reported GSR
<0.20	5
0.20–1.26	4–5
1.27–2.15	4
2.16–3.05	3–4
3.06–4.20	3
4.21–5.95	2–3
5.96–8.40	2
8.41–11.85	1–2
>11.85	1

11

The Quantification of Colour Appearance

We saw in chapter 7 that the only logical array of a random selection of colours is one based around a central vertical axis with black at the bottom and white at the top and that such an array, equally spaced in each of three dimensions, had been produced by Munsell. The configuration of XYZ space is, of course, quite different but that of $xy\ Y$ space is closer by having a central vertical achromatic axis which accounts for its popularity. Most of the colour difference equations developed up to 1976 used the distance element in a three-dimensional Euclidean space and one of the most important was the ANLAB space whose derivation is worthy of special mention.

The Adams chromatic value diagram

In 1923, Adams developed a colour vision theory in part based on Hering's theory but differing from it by postulating that the differencing mechanisms which produced the chromatic sensations were between the R and G cones to give the red/green response and between the B and G cones to give the yellow/blue response (Adams 1923). In 1942 he tested his theory against the Munsell colours. In the xy diagram colours of the same Munsell chroma plot as ellipses and the size of the ellipse for a given chroma is about four times larger at value 1 than at value 9. Adams reasoned that if his theory were correct, plotting functions of X–Y to represent R–G cone responses and Z–Y to represent B–G cone responses should result in colours of the same Munsell chroma plotting as circles of the same radius irrespective of the lightness. Before doing this, however, he made two important modifications.

The first modification was to overcome a defect of the xy diagram in quantifying appearance. A white surface appears white under D65 and A if the observer has become chromatically adapted to each illuminant in turn: chromatic adaptation will be considered in detail later in this chapter. The position of such a surface is not, however, the same as shown in figure 8.7. Adams therefore divided each tristimulus value by that of the illuminant which is perpetuated in the CIELAB equation by X/X_n, Y/Y_n and Z/Z_n. These ratios are identical for all achromatic colours and each will therefore be located in the same position irrespective of the illuminant used.

The second modification was to assume that the function which converted Y values into a uniformly spaced Munsell value scale, $V_y = Y^{0.5}$, should also be applied to X and Z values to give the chromatic values V_x and V_y. He then found that by plotting $(V_x - V_y)$ against $0.4 \times (V_z - V_y)$ colours of the same Munsell chroma did plot reasonably close to circles whose radii did not change much at different Y values (Adams 1942). The spacing of the Munsell colours was, in fact, so good that the infrequent inconsistencies were later used in the development of the Munsell renotations described in chapter 7.

ANLAB colour space

In 1944, Nickerson and Stultz modified the Adams chromatic values by using the Judd polynomial and introduced a correctly-scaled lightness co-ordinate, 0.23 Vy. The resulting colour difference equation was then scaled with a factor of 40 so as to make unit colour differences about the same size as the NBS unit (Nickerson and Stultz 1944).

In 1952 Glasser and Troy proposed changing $(V_z - V_y)$ to $(V_y - V_z)$ so that the hue order became the same as that in the xy diagram (Glasser and Troy 1952) and, later, the three components were given the symbols L, a and b following Hunter's earlier use of α and β (Hunter 1942). The space subsequently became widely known as ANLAB.

Though the subtractions $(V_X - V_Y)$ and $(V_Z - V_Y)$ were used by Adams because they are the mathematical representations of the inhibitory nerve connections (synapses) in the retina postulated in his theory of colour vision, when they are combined with V_Y they produce a colour space with exactly the same configuration as that of Munsell space. In chapter 10 we saw that the $L^*a^*b^*$ co-ordinates of CIELAB space were obtained by merely applying simpler functions to X/X_n, Y/Y_n and Z/Z_n. In consequence CIELAB space has the same configuration as Munsell space but two valuable advantages:

(*a*) the calculation of the co-ordinates is very much simpler;

(*b*) although the Munsell colours were equally spaced in each of three dimensions, the spacing in each dimension was different, i.e. one value step was visually equal to two chroma steps and to three hue steps at Chroma 5 (Nickerson 1936). The corresponding spacings in CIELAB space are more nearly equal.

The co-ordinates of Munsell space, however, are cylindrical, $L^*a^*b^*$ co-ordinates are Cartesian but they can be readily converted into the cylindrical co-ordinates L^*C^*h using the equations

$$C^*_{ab} = [(a^*)^2 + (b^*)^2]^{0.5}$$

$$h_{ab} = \arctan b^*/a^*$$

the latter being appropriately expressed on a 0–360° scale with the a^*+ axis being at 0° (360°) as shown in figure 11.1. The subscript ab was incorporated by the CIE to differentiate between CIELAB and CIELUV (L^* is common to both) but as CIELUV is rarely used for object colours it will be omitted for

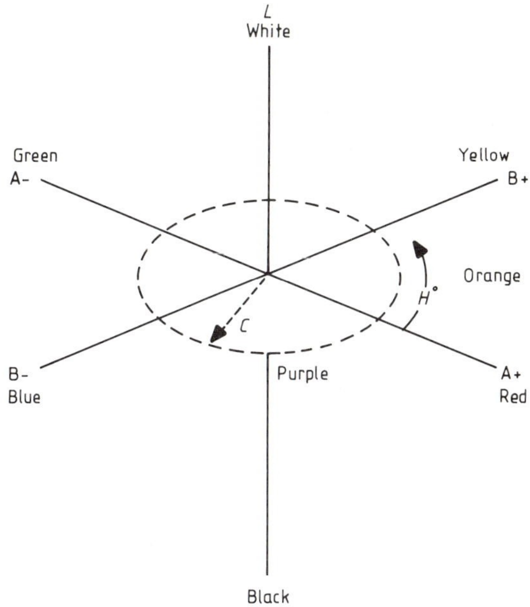

Figure 11.1 Cartesian and cylindrical co-ordinates of CIELAB space.

clarity. The asterisk is also often omitted for the same reason. Plate 2 illustrates the a^*b^* diagram which is destined to replace the xy chromaticity diagram (figure 8.7) in the case of object colours for many purposes. The colours represent those of maximum lightness at each position, i.e. it is a bird's-eye view of CIELAB space.

The most important use of these co-ordinates is in partitioning any colour difference into its three components and deriving the terms which would normally be used to describe them. The lightness difference is simply $L^*_{\mathrm{BATCH}} - L^*_{\mathrm{STANDARD}}$ and can be described as lighter if positive and darker if negative. The chroma difference is simply $C^*_{\mathrm{BATCH}} - C^*_{\mathrm{STANDARD}}$ and can be described as stronger if positive and weaker if negative. These definitions are included in the latest CIE vocabulary.

Quantifying and qualifying the hue difference component, however, is more complex and may account for the surprising fact that although *Lab* equations have been widely used in the USA since about 1950 the virtues of cylindrical co-ordinates were not appreciated until they became part of the CIELAB recommendation in 1976. The reason for this may date back to the pioneering days when Nickerson derived an index of fading based on the change in Munsell co-ordinates – ΔH, ΔV, ΔC. These differences, suitably scaled were then summed (Nickerson 1936). This equation, however, is based on a complicated non-Euclidean geometry but in 1951 Godlove derived a method of using the Euclidean distance element expressed in cylindrical co-ordinates (Godlove 1951): this was the Munsell renotation formula recommended for study by the

CIE in 1967. In 1951, however, the conversion of xyY values into HVC involved interpolations in three dimensions which made it quite impracticable. This complication, however, could be avoided by 'steering clear of cylindrical co-ordinates' (Judd and Wyszecki 1975, p. 316) and this is exactly what happened.

Godlove introduced a cosine term to allow for the fact that a given difference in hue angle represents a much larger perceived hue difference for high chroma colours than for low chroma colours. A much simpler method was, however, devised independently in Germany and in the UK in the early 1970s when ANLAB was being developed as a national standard. The combined hue and chroma difference, H/C, is given by

$$(\Delta H/C)^2=(\Delta E)^2-(\Delta L)^2.$$

As ΔC can also be calculated $(C_{\mathrm{BATCH}}-C_{\mathrm{STANDARD}})$ these two can be represented as the hypotenuse and one side of a right-angled triangle. The remaining side then represents the hue difference, ΔH, in the same units and independent of the value of C

$$(\Delta H)^2=(\Delta H/C)^2-(\Delta C)^2$$

(Strocka 1971).

Alternatively

$$(\Delta H)^2=(\Delta E)^2-(\Delta L)^2-(\Delta C)^2$$

(Cooper and McLaren 1973a), this version becoming part of the CIELAB and CIELUV equations.

Hue differences can be qualified by rotating the chroma axis which passes through the position of standard towards the position of batch, continuing the rotation, if necessary, until it crosses one of the four axes. The hue associated with this axis will often indicate the nature of any hue difference. This method fails, however, whenever the hue of standard is the same as the indicated hue difference since to describe, for example, a batch as being redder than a red standard is meaningless. This weakness was overcome by Taylor who suggested always continuing the rotation until the second axis is crossed and putting the hue associated with this axis in brackets: one of the terms is always meaningful except for achromatic colours (McLaren and Taylor 1981).

The failure of this method for achromatic colours is mainly due to the fact that an increase in chroma is invariably perceived as a difference in hue. It is further complicated by the fact that a grey where $a^*=b^*=0$, i.e. which is colorimetrically neutral, is not perceived as neutral by the majority of colourists. Their choice lies between the blue and green axes of the a^*b^* diagram.

The method of partitioning used in the USA before the value of cylindrical co-ordinates was appreciated was to report ΔL, Δa and Δb. The last two were often accompanied by hue difference terms

$a+$ redder	$b+$ yellower
$a-$ greener	$b-$ bluer

but this tended to imply that the only possible difference other than lightness was that of hue. Even if there was no hue difference but a chroma difference, this would be indicated by one or usually two hue difference terms.

The dyer's variables of perceived colour

Whereas the variables of lightness, chroma and hue are natural variables in the sense that they can be deduced introspectively by viewing coloured surfaces without any knowledge of how the coloured effects have been produced, the dyer's variables are based on the effect on appearance caused by variations in dyeing, the most important being the amount of dye applied to the fibre. Increasing the amount of a single dye (or a compatible mixture) causes an increase in *depth* and the depth lines of a representative selection of dyes are shown in the L^*/C^* diagram in figure 11.2(*a*) and in the h/C^* diagram in figure 11.2(*b*). The dyer's most important variable therefore affects all three Munsell variables. The terms normally used by dyers to describe differences in appearance which can be eliminated merely by altering the amount of dye on the fibre are *fuller* if more dye appears to be present and *thinner* if less appears to be present (Morton 1976).

The dyer's second variable, that of *hue*, is quite straightforward: the addition of a smaller amount of a shading dye causes mainly a change in hue angle and the dyer normally describes such a change as redder, yellower, greener or bluer.

If two dyeings are judged to be equal in depth and hue but do not match they differ with respect to the dyer's third variable, that of *brightness* which can be most readily defined by defining its opposite quality. If a trace of a black dye is added to any chromatic dyeing this makes it *flatter*, the original dyeing being described as *brighter*. The addition of a trace of the dye of complementary hue will have the same effect, of course.

The locus of dyeings containing increasing amounts of added black is called the *dullness* line and duller is sometimes used as synonym for flatter. Typical dullness lines are shown in figure 11.3 (McLaren 1972, Cooper and McLaren 1973a).

It will be readily seen that, strictly speaking, the terms flatter and brighter should not be applied to grey or black dyeings because the effects will be identical with fuller and thinner. Brighter is, however, a term widely applied to greys and blacks and seems to be synonymous with blueness (McLaren 1973b).

It is obvious that the colorimetric differences lighter/darker, weaker/stronger can only be converted into the dyer's differences fuller/thinner, brighter/flatter if one locates the position of standard in an L^*C^* diagram and the direction of the depth line: it is then reasonable to assume that the dullness line lies in the direction of $L^* = C^* = 0$. A computer program for splitting any colour difference into the dyer's variables of hue, brightness and depth was written by Cooper in 1973 and is part of the M&S 83A colour difference formula.

The paint technologist's variables of perceived colour

The paint technologist's variables are also based on the visual effects caused by changes in pigment formulation and the major difference from the dyer's

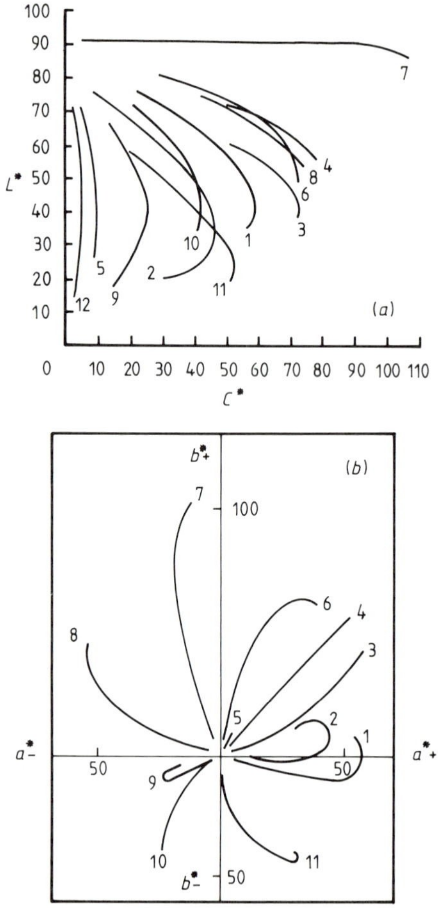

Figure 11.2 Depth lines of typical dyes. (1) CI Reactive Red 11; (2) CI Direct Red 1; (3) CI Acid Red 85; (4) CI Reactive Orange 13; (5) CI Vat Brown 49; (6) CI Disperse Orange 13; (7) CI Reactive Yellow 86; (8) CI Reactive Blue 80; (9) CI Acid Green 25:1; (10) CI Disperse Blue 87; (11) CI Acid Violet 41; (12) CI Disperse Blue 35.

variables arises from the fact that white is also a colorant. If paint A contains more white pigment than B, A will be described as *whiter* than B and, conversely, B will be described as *deeper* than A. The paint technologist also recognises the dyer's variable of brightness but his terminology is *cleaner* and *dirtier*. The presence of more or less chromatic pigment is described as *stronger* or *weaker*.

Again the conversion of the colorimetric differences into the paint technologist's variables can only be done by plotting in an L^*C^* diagram and it is further

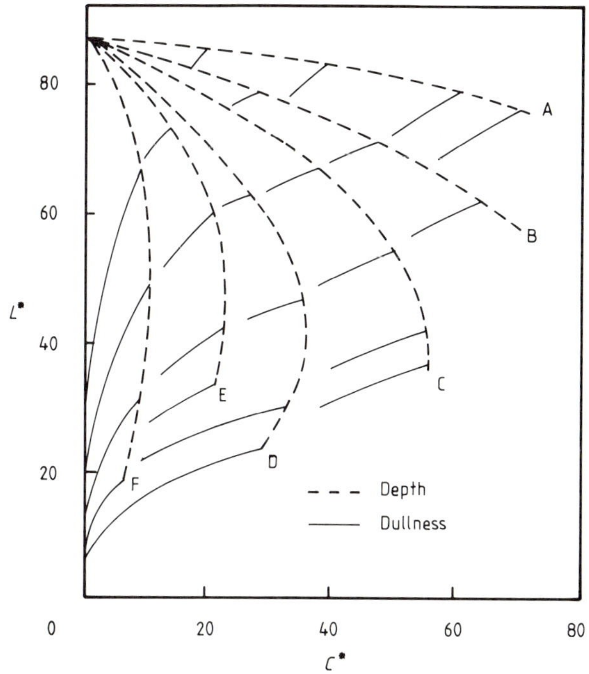

Figure 11.3 Depth and dullness lines of typical dyes. (A) CI Reactive Yellow 2; (B) CI Reactive Orange 2; (C) CI Reactive Red 3; (D) CI Reactive Violet 1; (E) CI Reactive Green 44; (F) CI Reactive Black 4.

complicated by the fact that all inorganic and some organic pigments give depth lines like those in figure 11.4(*a*) whilst most organic pigments give depth lines like those in figure 11.4(*b*). The cause of the difference must lie in the spectral refractive index of the pigment.

The quantification of whiteness

The usefulness of a colorimetric method for quantifying whiteness has been recognised since 1934 when the first whiteness formula was published (MacAdam 1934). Since then there have been many more – whiteness formulae are much more numerous than colour difference formulae – and in 1967 the CIE began an evaluation of the most important. This study turned out to be extremely protracted as it was not until 1982 that the CIE whiteness index was devised

$$W = Y + 800(x_n - x) + 1700(y_n - y)$$

where x, y and Y are the colorimetric values for the specimen under D65 (2° or 10° observer) and x_n, y_n the chromaticity co-ordinates of the light source.

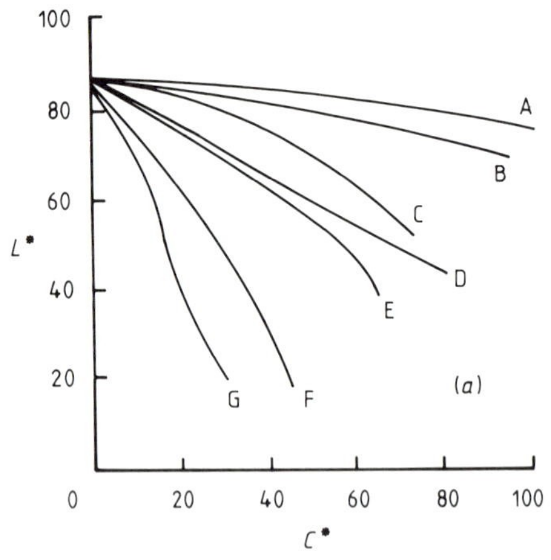

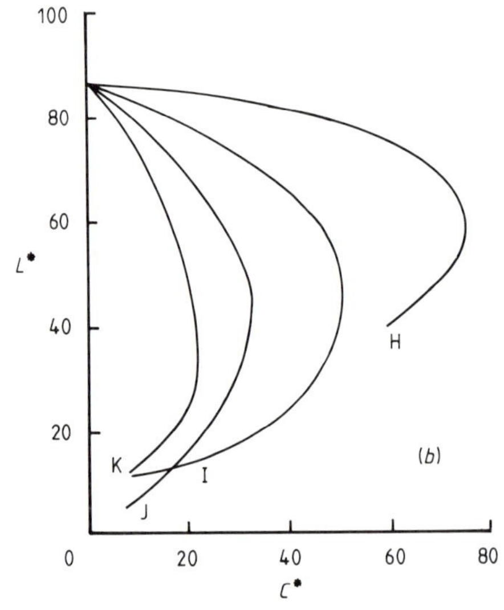

Figure 11.4 Depth lines of typical pigments. (A) CI Pigment Yellow 34; (B) CI Pigment Orange 1; (C) CI Pigment Orange 5; (D) CI Pigment Red 104; (E) CI Pigment Green 13; (F) CI Pigment Red 12; (G) CI Pigment Brown 6; (H) CI Pigment Yellow 24; (I) CI Pigment Green 41; (J) CI Pigment Blue 27; (K) CI Pigment Violet 23.

The perfect reflecting diffuser has a whiteness index of 100.00, materials treated with a fluorescent brightening agent giving values as high as 150. When increasing amounts of such agents are applied to a white substrate, its appearance becomes whiter. The lightness does not increase to any extent but the chroma does – in the direction of blue. However, this blueness only becomes apparent if the material, illuminated by daylight, is seen against fresh snow, for example. The specific hue imparted by the agent is important and the index is accompanied by a tint factor

$$T_W = 1000(x_n - x) - 650(y_n - y)$$

$$T_{W,10} = 900(x_{n,10} - x_{10}) - 650(y_{n,10} - y_{10}).$$

If T_W is near zero, the specimen has a dominant wavelength of 466 nm. The more positive the value of T the greater the indicated green-ness: the more negative, the greater is the indicated redness.

External factors affecting colour appearance

The CIE system of colour measurement is based entirely on colour matches made against a dark surround, the so-called *Maxwellian view*. Fortunately extensive applications of the results of these experiments for over half a century have shown that they are valid for quantifying the colours of the great majority of objects seen under the quite different surroundings occurring in everyday life. But there are two situations where the perceived colour is not reliably quantified by its tristimulus values. The first is known as *simultaneous contrast* whose effects can only be assessed visually. The second involves *chromatic adaptation* but in this case extensive studies have resulted in the CIE system being extended to cover these effects with a fair degree of reliability.

Simultaneous contrast

Every designer of textiles, wallpapers etc knows that the colour of any particular area is markedly affected by the colour of any adjacent area. This was effectively exploited in the Gobelin tapestries by Chevreul who wrote a book on the subject in 1839. The most effective display of this phenomenon was a piece of carpet designed by Sidney Harry of the Regional College of Art, Bradford, which he showed at the first conference of the AIC (International Colour Association) held in Stockholm in 1969 (Harry 1969). Twelve colours were easily recognised in the geometrical design but the carpet was a Wilton carpet which meant that only five yarns were used, the other seven colours resulted from simultaneous contrast.

The best known failure of a designer to recognise the phenomenon involved the ceremonial sword presented by the British government to the people of Stalingrad in recognition of the heroic stand made by the city in the Second World War. The handle was covered with silk thread dyed to a brilliant turquoise shade and overlaid by gold wire. This had a decided pink colour reminiscent of ashtrays made of dyed anodised aluminium which were in widespread use at the time (Warburton and Lund 1956). The effect of simultaneous contrast is that the larger area induces its complementary colour into the colour of the adjacent smaller area. The complementary colour of turquoise is red and hence the gold wire appeared pink.

An identical effect occurs with the appearance of an *after-image*. After an observer has fixed his gaze on a coloured display for 15 seconds or more and then looks at a plain white field he sees an after-image in the complementary colours which persists for some time. One popular display uses a slide of a green elephant, the after-image having an alcoholic connotation, but a more impressive one uses the design of the Union Jack or the Stars and Stripes but in cyan, black and yellow. Surgeons' gowns are now green so that the appearance of the after-image of blood is not obtrusive.

The generally accepted explanation of the appearance of an after-image is that prolonged exposure to light of any one colour results in significant bleaching of the rhodopsin in the cones which are most sensitive to light of this colour. Regeneration via the intermediates listed in figure 5.1 is a slow process so initially the eye is less sensitive to light of this colour and if the observer looks at a white or grey area the complementary colour is perceived.

However, although the effects are identical, this cannot explain the phenomenon of simultaneous contrast because in such cases the effects are virtually instantaneous – less than 0.1 s and the cause must lie in interactions between the signals sent from the cones to the visual cortex.

Chromatic adaptation

Every colourist is well aware of the fact that if a single specimen is being viewed in a colour matching cabinet and the light source is suddenly switched, for example, from a daylight simulator to tungsten, virtually all colours appear to have changed. At the same time, however, the colourist is perfectly aware that in most cases the immediate change in perceived colour does not appear to apply to the object but to the light source. This immediate colour change is quantified by the change in chromaticity co-ordinates and is known as the 'colorimetric colour shift'. An example is given in figure 8.7 where the points marked D65 and A represent the colorimetric colour shift for a card which appears white in daylight and yellow when immediately viewed under tungsten light. If the observer takes the card into a room lit solely by the light from the same tungsten filament lamps, after a few minutes it will appear white again. This reversion to the colour perceived under the first light is known as 'chromatic adaptation'. With one exception, natural coloured objects – animal, vegetable and mineral – revert reasonably closely to the colours they had in daylight and are said to possess 'colour constancy' the exception being the rare mineral *alexandrite* which appears green in daylight but red under tungsten light even after complete chromatic adaptation.

The phenomenon of chromatic adaptation is one of the most remarkable characteristics of colour vision and it is easy to see *why* it evolved in animals using the sense of colour in food and predator recognition. Colour identification would have to be reliable under the reddish light of sunrise and sunset and also under the light from a blue sky falling on objects in shadow whose spectral power distribution is quite different.

The SPD of tungsten filament lamps falls between these extremes and it is not surprising that natural objects do possess reasonable colour constancy under this light. Objects coloured by mixtures of colorants, particularly dyes, may exhibit inadequate colour constancy if the choice of colorants has not been a good one. Dyemakers therefore often include information about the change of shade in artificial light so that this can be avoided by not using mixtures of dyes which change in the same direction. These changes are assessed visually and are of the colorimetric colour shift, not the residual colour shift which is what is really required. Residual colour shifts are of particular importance in the photographic industry where prints and transparencies are often viewed in daylight and in tungsten light. Any marked lack of colour constancy will be unacceptable particularly in flesh tones and the need to be able to quantify this characteristic is one of the reasons why there is now 'an amount of experimental and theoretical information bearing on the subject so vast that it would fill an entire book even in cryptic summary form' (Bartleson 1978). The most recent method for calculating the effects of chromatic adaptation between daylight and tungsten light is that of Hunt and Pointer (1985).

The most widely used method of calculating the effects of chromatic adaptation – it is used in the CIE method for determining the colour rendering properties of light sources (CIE 1974) – is based on the simple theory put forward by Helmholtz who said 'that in our observations with the sense of vision we always start out by forming a judgement about the colours of bodies, eliminating the differences of illumination by which a body is revealed to us . . . the acts of judgement . . . are always executed unconsciously and involuntarily' (Helmholtz 1924). This theory became generally known as the 'discounting the illuminant colour' theory and was not seriously challenged until the *retinex* theory, which will be considered later, was proposed in 1964. Von Kries, in 1905, suggested that when the observer becomes chromatically adapted to a second illuminant the sensitivity of each of the cone mechanisms is altered by a constant factor. The values of these factors are determined by the changes in SPD of the illuminant. Tungsten light being richer in red and deficient in blue causes a decrease in the sensitivity of the red cones and an increase in that of the blue cones. These factors, known as the von Kries coefficients, are then used to calculate tristimulus values under the second illuminant:

$$R_2 = \alpha R_1$$

$$G_2 = \beta G_1$$

$$V_2 = \gamma V_1$$

where R, G and V refer to the fundamental primaries of the trichromatic theory which can be expressed in terms of X, Y and Z (Judd 1945).

The von Kries coefficients are calculated from the tristimulus values under both illuminants of a specimen known to possess perfect colour constancy and the one invariably taken is that of the perfect reflecting diffuser. This has the same tristimulus as the light source, X_n, Y_n, Z_n, and as these values are involved in the CIELAB formula this has its own in-built von Kries transformation: every illuminant, for example, has the values $a^* = b^* = 0$ indicating perfect colour constancy for the perfect reflecting diffuser. CIELAB, however, does not involve the use of the fundamental primaries but the effect of this has been found to be negligible for quantifying colour constancy in the colorant-using industries. Here the only important aspect is if the perceived hue under the second illuminant is significantly different from that which the observer remembers seeing in daylight. This can be reliably assessed by calculating the CIELAB hue angle, *h*, under any illuminant and is illustrated by the three dark brown shades, devised by Carbone (1962) and dyed on wool with

(1)		(2)		(3)	
	Acid Red 1		Acid Red 1		Acid Red 1
	Acid Yellow 17		Acid Yellow 17		Acid Orange 10
	Acid Blue 45		Acid Orange 10		Acid Blue 45
	Acid Green 16		Acid Blue 45		Acid Blue 47

whose reflectance curves are shown in figure 11.5. Under tunsten light even the colorimetric colour shift is only slight but after chromatic adaptation to the

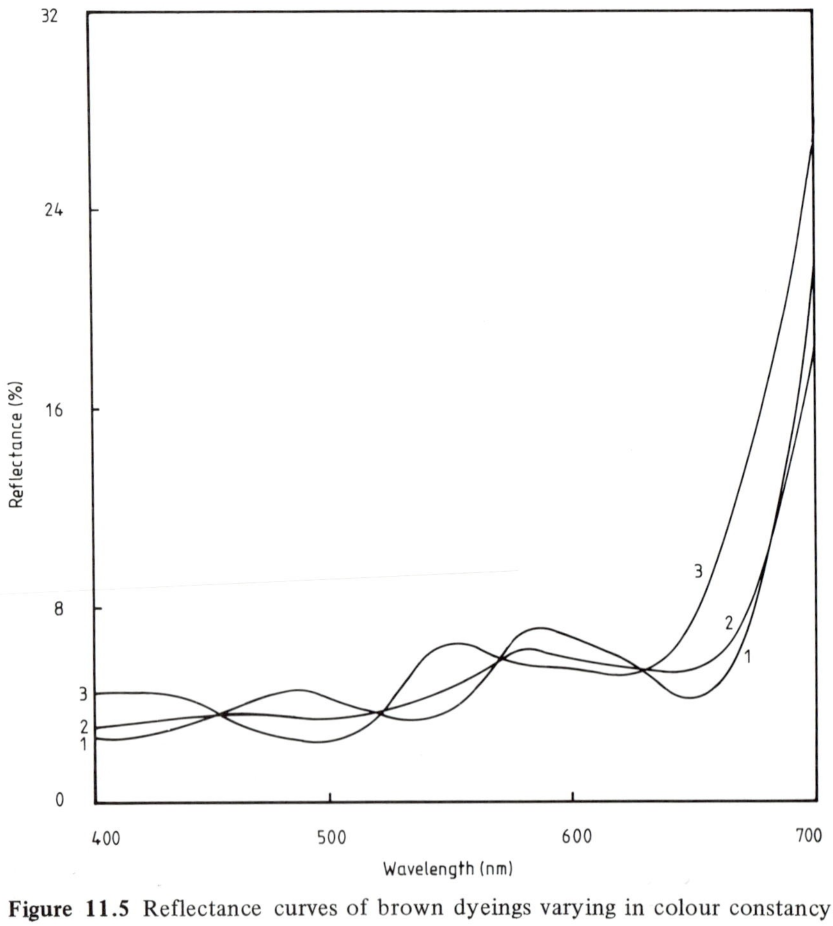

Figure 11.5 Reflectance curves of brown dyeings varying in colour constancy.

Philips three-band lamp, Colour 84, only No. 2 would be described as having the same colour as in daylight while No. 1 would be described as being a much redder shade of brown and No. 3 is unmistakably olive green.

These effects are reliably quantified by the CIELAB hue angles under the three illuminants as shown in table 11.1.

Table 11.1 CIELAB hue angles of three brown dyeings under three illuminants.

	Dyeing number		
Illuminant	1	2	3
D65	52	61	71
A	52	60	69
Colour 84	40	60	100

Recently, however, an instrumental method of assessing colour constancy, based on the $\Delta E(Mc)^2$ formula has been proposed (Berns *et al* 1985).

Specimens illustrating very poor colour constancy can be obtained by dyeing blue and green shades with mixtures of CI Acid Violet 17 and CI Acid Yellow 3: under tungsten light the blue shades appear purple and the green shades, brown.

The best specimen for illustrating lack of colour constancy

All of the dyed textiles and pigmented cards which have been used to demonstrate a lack of colour constancy show marked hue changes in one direction only. If a specimen is made by placing a didymium glass filter (mentioned in chapter 2) over a piece of white paper it will exhibit the following colours under different illuminants

daylight	bluish grey
tungsten	pink
Colour 84	green.

Edwin H Land's contributions to colour science

The two-colour projections

In 1955, Land, the inventor of 'Polaroid' and of the instant black and white camera, made a remarkable demonstration of multicoloured images produced by two projectors, one normal, the other having a red filter over the lens. The slides were separation positives taken through green and red filters respectively of a table-top collection of familiar objects. When the two images were out of register all that appeared on the screen were patches of white, pink and red lights but when they were brought into register the objects were seen to have their correct colours (Land 1959a,b). Anyone who witnessed one of these demonstrations will agree that they were extremely impressive as colours other than red, pink and white were entirely unexpected except to a handful of experts or to those familiar with the development of cinematography in colour. From 1906 until 1928 this was invariably achieved by projecting separation positives through red and green filters but in 1914 Fox and Hickey discovered that although the removal of the green filter reduced the colour gamut this was adequately compensated by the increase in brightness of the image. This was first commercialised in 1928 as the *Raycol Process* (Cornwell-Clyne 1951).

Unfortunately, however, Land's demonstrations were described in the magazine *Fortune* by a reporter who stated that 'a theory of colour that has stood for nearly 300 years has suddenly been overthrown ...' and 'scientists since Newton have been completely fooled about the way the eye sees colour' so that 'every text-book dealing with colour will have to be re-written' (Bello 1959). This was not entirely an example of journalistic exaggeration for although in his article in the *Scientific American* Land said that the 'long line of great investigators cannot have been mistaken' he went on to say 'that their work had very little to do with colour as we normally see it. They dealt with spots of light and particularly with pairs of spots trying to match one to another' (Land 1959c). In his paper in the *Proceedings of the National Academy of Science* he also stated that 'color in images cannot be described in terms of wavelength' (Land 1959a).

The article in *Fortune* had a marked impact on television companies in the USA as it suggested that colour television could be transmitted by two channels rather than three. McCamy of the National Bureau of Standards then made over 50 demonstrations of the Land two-colour projections throughout the USA which confirmed that a surprisingly large gamut of colour was displayed but he ended his demonstration by showing a Kodachrome transparency of the same scene projected at the same brightness (McCamy 1960). The lack of saturation and of subtle differences in shade in the two-colour projections was so marked that, of course, nothing further was heard about two-channel colour television.

This phenomenon is an elaborate extension of the appearance of coloured shadows first observed by Leonardo da Vinci in the fifteenth century and extensively studied by Goethe in the nineteenth. A simple demonstration of this impressive phenomenon used to be described in many elementary textbooks on light. A red and a white projector with no slides in the carriers shines on to a screen which appears pinkish-white in colour. If any opaque object is now placed between the projectors and the screen, two shadows appear, one lit only by red light which appears red as one would expect. The other is lit only by white light but unexpectedly appears blue-green.

This phenomenon is easily explained by the Helmholtz 'discounting the illuminant colour' theory of chromatic adaptation. The pinkish-white colour of the screen is assumed to be that of an illuminant and is automatically discounted so that it appears white. The shadow that is lit by the light coming from the projector with no filter is less red than the screen and must therefore appear blue-green. Whatever the colour of the projected lights – and one does not have to be white – the colours of the shadows are always complementary.

It is of no significance that the colours perceived were those of the original objects as the same colours were produced by Brocklebank, Karp and Wilson using neutral density filters to provide 36 different red-to-white ratios on the screen from 0 : 0 to 1 : 1. When the squares were adjacent or separated by a black grid only the expected red/white mixture colours were perceived but when each was separated by a clear area the whole colour gamut appeared. Now the clear areas provided a pinkish-white background as a reference white for discounting the colour of the illuminant and it is significant that such an array does not appear to be patches of light but appears to be a collection of coloured papers on a white board. Viewing any square through a narrow tube to isolate the area destroyed all the colours other than red/white mixtures (Karp 1959, Wilson and Brocklebank 1960).

Land had, however, specifically rejected chromatic adaptation as being the cause of the appearance of unexpected colours because they appeared instantaneously and he had deduced from the standard textbook, *The Science of Color* (Committee on Colorimetry 1953), that 'adaptation implies that the colors in our images are the result of extended exposure of the observer's eye to the image' (Land 1959d). He had, however, overlooked the statement on page 112 that the 'most striking temporal feature of light adaptation is the suddenness and extent of the initial phase'. This referred to the studies made by Schouten and Ornstein which showed that the initial phase is complete in about 0.2 s and nearly so in 0.05 s (Schouten and Ornstein 1939). They termed this initial phase α adaptation and concluded that its origin must lie in the retinal synapses. The more familiar process, which Walls characteristically described as being 'regulated by the cold-molasses kinetics of photopigment-concentration' (Walls 1960), they termed β adaptation.

Land's papers were the subject of a brilliant fourteen-page review by Judd (1960) who showed that Land's conclusion that colour in images cannot be described in terms of wavelength was wrong and that all of the colours could be predicted by the empirical formulae he had developed in 1940 (Judd 1940). Although the review was mainly critical it did include the comment that 'It is nevertheless remarkable that in a few years of intensive study Land, and his associates, should have been able to rediscover independently so large a fraction of the known phenomena of object-color perception, and it is not too surprising that he should have been led by them to the false hypothesis that color in images cannot be described in terms of wavelength. The literature of object-color perception is studded with hypotheses that have later been found to be unacceptable.' Other reviews were published by Wilson and Brocklebank (1961), by Hurvich and Jameson (1960) and by Walls (1960) the tenor of the last-named being indicated by its title 'Land!, Land!'

The Mondrian experiments

To eliminate the possibility that the observer could be discounting the colour of the illuminant and to prevent perceived colours being induced by either simultaneous or successive contrast Land mounted about 100 rectangles of papers of many different colours and sizes on a white board. The surfaces were matt as specularly reflected light has the same colour as the illuminant. To eliminate any background being visible the rectangles overlapped. The display was termed a 'Mondrian' as it resembled paintings of the Dutch artist and it also eliminated any possibility of an observer being influenced by knowing what the true colour of any area was. It was illuminated by independent red, green and blue projectors and the amounts of red, green and blue light reflected from any area could be measured with a telescopic photometer (Land 1977, 1983). From the very large number of experiments carried out by Land and his colleagues the *experimentum crucis* (to use Newton's phrase) is as follows. The beams from the three projectors were restricted by masks to one area of *any* colour other than white or grey and the observer, sitting in a completely darkened room with no stray light from the projectors entering the eye, balances the amounts of *RGB* to make the patch of light appear white. The amounts of red, green and blue light entering the observer's eye are measured. These are the *RGB* tristimulus values of that area seen as white in Maxwellian view, i.e. that used to obtain the CIE colour matching functions. The three projectors are then switched off and the masks removed. When they are switched on, although the composition of the light from that area is unchanged, it no longer appears white but the colour it has in daylight. Other experiments showed that the composition of the illuminant could be varied over a very wide range but the perceived colour of each area remained reasonably constant.

The degree of colour constancy was later quantified (McCann *et al* 1976) by having the observers view a Mondrian with one eye and match it with Munsell papers illuminated by a constant white light viewed by the other eye. A Mondrian containing 17 colours was used under five mixtures of red, green and blue lights. The first mixture was such as to appear a good white and the amounts of *R*, *G* and *B* reflected from a mid-grey area in the Mondrian were measured. The other mixtures were such as to make the amounts of *R*, *G* and *B* reflected from red, yellow and blue areas exactly the same as were reflected from the grey area. Each of these areas would therefore appear white under its respective mixture if viewed in isolation. Although the degree of colour constancy was generally good there were exceptions, the worst being the white and mid-grey areas whose perceived colours corresponded to the colour of the illuminant, in some cases as intense as Chroma 4.

Land claimed that when synchronised shutters were placed on the projectors and the Mondrian was illuminated for only 0.1 s 'there is an even closer approximation to independence of color from the composition of the radiation from an area to the eye' (Land 1983). If this statement means that continued observation of the Mondrian causes the perceived colour of each area to shift further away from its daylight colour it is contrary to common experience. After coloured sunglasses have been worn for some time the wearer is completely unaware that they are coloured unless the outside world is glimpsed around the frame. Land's claim was not based on the experiments of McCann *et al* (1976) as their observers were, unfortunately, given unlimited time to make their matches and these times were not recorded: but this is not important.

What is highly significant is Land's discovery that there is a near-instantaneous mechanism causing approximate colour constancy which cannot be explained by the 130-year old theory of discounting the colour of the illuminant because the observing conditions were such that there was no possibility whatsoever of the observer being able to perceive the illuminant. In a lecture given by Land in 1958 (unpublished) he suggested that the observer might be using the average chromaticity of the whole field of view as a reference point, a view which was also held by Judd (1959). After the Mondrian experiments were carried out Land developed this idea into the first comprehensive theory explaining colour constancy (Land 1964). As this involved neural signal comparison which could occur in either the retina or the cortex or in both, Land coined the term *retinex* to replace the cumbersome 'eye/brain' alternative.

The retinex theory

If one views a scene through a red filter and the intensity of the illuminant is changed one of the characteristics which remains constant is the lightness of a given object relative to all of the others. The same is true if the experiment is repeated through green and blue filters. Land therefore suggested that the retinex might assess relative lightnesses for each of the red, green and blue channels involved in colour vision independently. The perceived colour of an object was not directly governed by the relative strengths of the R, G and B signals coming from the cones but by the three relative lightnesses of the object. To achieve this the retinex would therefore have to be able to compare the strength of the signal coming from the centre of the receptive field with that coming from the surround for each channel independently and then to compare the three resultant signals to deduce the perceived colour.

In order to determine whether this neural computation occurred in the retina or in the cortex Land *et al* recorded the perceived colour of a test area in a Mondrian viewed by an observer whose corpus callosum had been severed. The corpus callosum is the linkage between the two halves of the brain and is frequently severed in subjects suffering from severe epileptic seizures. Although this does not affect normal vision, if an object is seen only by the right hemisphere the observer cannot describe it because the left hemisphere is solely responsible for speech. The surrounding Mondrian was viewed alternately in the left and right visual fields and from the colour of the test area in the centre of the visual field perceived by the subject Land *et al* deduced that 'the calculation could not have occurred solely or largely in the retina and must have occurred in the cortex' (Land *et al* 1983).

In another equally ingenious experiment Ingle trained goldfish to recognise specific colours in a Mondrian illuminated by daylight. When the composition of the illuminant was markedly changed the goldfish still recognised these areas showing that colour constancy was a characteristic of an animal with only a rudimentary brain (Ingle 1983). This characteristic had been discovered earlier in fishes (Burkamp 1923) and hens (Katz and Révész 1921).

If the retinex theory is correct there must be neurons capable of comparing the strength of signals from the centre of the receptive field with some kind of average of those coming from the surrounding area. The spatially opponent/spectrally non-opponent cells described in chapter 5 are likely candidates for computing overall lightness but no spatially opponent cells activated by a single cone type i.e. $R+/R-$, $G+/G-$ and $B+/B-$ have ever been found. Marr (1982) however has suggested that the double opponent cells described in chapter 5 might be the initial step in neural computation.

The method of computing relative lightnesses devised by Land and McCann (1971) takes a random path across the Mondrian, differentiates log luminance along the path, sets all small values to zero (which removes the effects of slow changes in luminance) and re-integrates to obtain the lightness distribution along that path. When enough random paths have been chosen to cover the whole Mondrian, the computation is complete (Marr 1974). This has since been modified (McCann *et al* 1976) and has recently been termed the Reset Ratio-Product model (McCann and Houston 1983). Whether the complicated mathematics involved in the method could ever be carried out by neurophysiological methods of computation must be an open question.

The impact of Land's discoveries

(*a*) *On the practice of colorimetry.* In both the two-colour projection demonstrations and those involving Mondrians Land has clearly demonstrated that the perceived colour of an area is not governed solely by the spectral composition of the light entering the eye unless this is the only source of retinal excitation. He therefore concluded that 'only our eyes can categorize the color of objects, spectrophotometers cannot' (Land 1977) a similar view being expressed by Zeki – 'the machine, unlike us, is equipped to measure wavelengths alone; the human visual system sees only colours' (Zeki 1984). In 1960 Judd's main criticism of Land's oral and written presentations was that they lacked 'any serious attempt to relate the results of his experiments to previous studies of the same phenomena' (Judd 1960). Land's 1977 statement invites the same criticism because colorimetrists have not only known that *XYZ* tristimulus values do not reliably quantify colour appearance even under tungsten light, they have been allowing for this by applying the von Kries transformation since 1952. The only mention of this by Land and his colleagues was in a footnote to the paper by McCann *et al* which promised to compare the reliability of the retinex method of calculating colour co-ordinates with those involving chromatic adaptation transformations (McCann *et al* 1976); this has not yet appeared. If their method should be superior then it will be of very limited interest as at present it is only applicable to the colours in a Mondrian array and then it requires the measurement of the amounts of red, green and blue light reflected from 'as few (*sic*) as one or two hundred' areas lit by the specific illuminant (Land 1983).

(*b*) *On colour vision theory.* If electrophysiological studies confirm that one of the stages in the visual process is the comparison of the lightness of the area in the centre of the receptive field with some average of the surrounding area for each of the *RGB* channels this can be easily incorporated into current colour vision theory which is far from completion. It will be no more difficult than the incorporation of the opponent-colours theory into the long-accepted trichromatic theory following the discovery in 1953 of cells giving opponent-colours responses. This was rapidly accepted as explaining a stage in signal processing which occurs between the cones and the cortex and the comparison of reflectances could be another stage either earlier or later. Its acceptance would not necessarily invalidate the generally accepted theory that slow chromatic adaptation is caused by photochemical bleaching though Schouten and Ornstein (1939) concluded that this β adaptation was also due to 'electrical processes' in the retina.

Newton's theory that 'Every Body reflects the Rays of its own Colour more copiously than the rest' (Newton 1730) will still be true though the rest of the sentence 'and from their excess and predominance in the *reflected* Light has its Colour' (my italics) is incorrect even if Land's theory should be disproved. This, however, by no means justifies Bello's claim that 'every text-book dealing with color will have to be re-written' nor that part of the 'Horizon' programme on BBC television (January 7 1985) which stated that Land had put forward a 'new set of ideas that completely overturn our old theories about how we see colours'. The presenter did, however, say that these ideas 'seem so improbable to the majority of scientists working on colour vision that they've simply ignored them'. The latter part of this statement is undeniably true and the reasons were given in an article by one of the participants in the programme. This, however, ended more optimistically with the words 'The establishment and Land are coming together' (Mollon 1985, 1986).

12

Computer Match Prediction

> A symposium on colour held in the USA in 1940 was attended by about 150 paint, pigment, and dye men, both executives and technologists, illuminating engineers, and plastic experts. Four nationally known speakers regaled this audience with 2 hours of spectrophotometry, CIE system, colour standards, effect of illuminant on object color, photoelectric colorimetry and so on. The audience became drowsy, with here and there someone frankly snoring. At the end of the symposium, the chairman called for questions. There were a few polite queries by the more wakeful, but most of the audience remained in a state of suspended animation, and soon the discussion died away practically to zero. The chairman insisted 'Now that you have the experts at your disposal', he said, 'this is your chance to find out what you want to know about color'. Thus spurred, a paint man got to his feet and said, 'What I want is a photoelectric colorimeter that will read the proportions of the pigments required to produce a color match for a specimen'. Like magic the somnolence disappeared. The audience snapped to attention and the ensuing discussion was followed both eagerly and intelligently†.

Two years later a method of doing this (though not with a colorimeter) was developed at the UK Paint Research Station (Duncan 1963) although the method was not published until 1949 (Duncan 1949). It was, however, completely impractical as each calculation took several hours and match prediction only became a practicable proposition when computers became available in the late 1950s. Since then it has become the most important industrial application of colorimetry. Because it is also the most complex, no attempt will be made to cover the subject in depth but the basic principle will be described together with an historical account of the key developments. Other reviews have been published by Brockes (1965), Gall (1973), Kuehni (1975) and Best (1981).

Figures 12.1-12.3 are reflectance curves of dyeings made with increasing amounts of typical red, yellow and blue dyes: they also illustrate the general shapes of the curves of white paints, whose reflectance curves are similar to those marked 'substrates' to which have been added increasing amounts of red,

† Quoted from Judd and Wyszecki 1975, p. 446 by kind permission of John Wiley and Sons Inc.

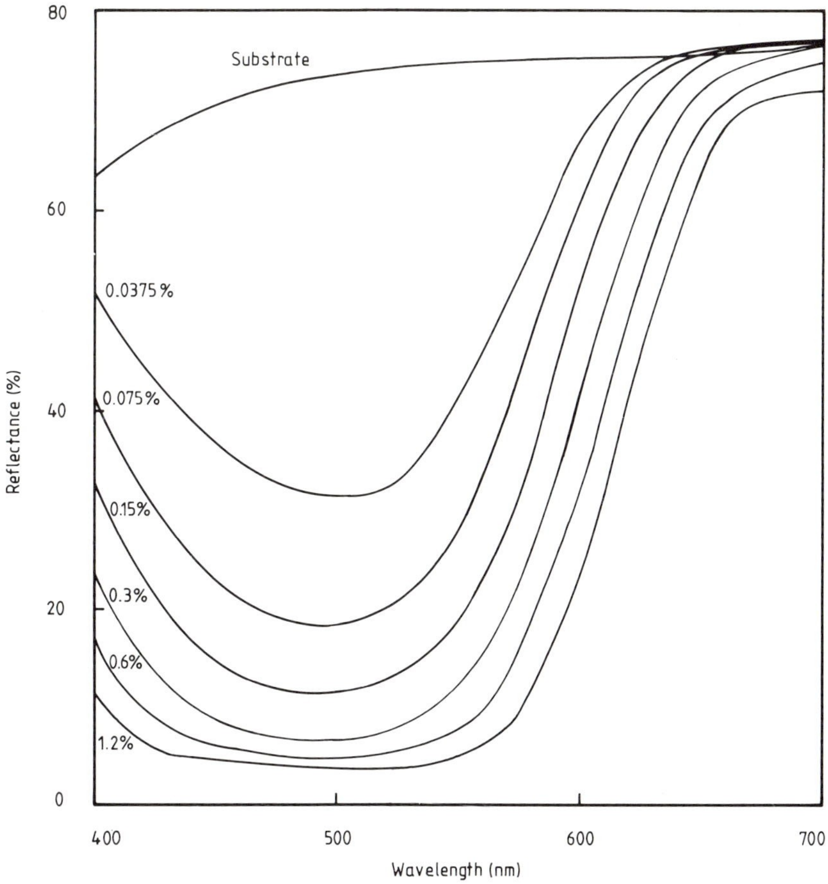

Figure 12.1 Reflectance curves of increasing depths of red dye.

yellow and blue tinters. Figure 12.4 is the reflectance curve of a dyeing made with a mixture of the three dyes and if it were possible to predict the reflectance curve from a knowledge of the amounts of the components used, this would form the simplest basis for the inverse process – match prediction.

The change in reflectance with increasing concentration is clearly wavelength-dependent and at any wavelength the relationship between the two is not simple. One therefore requires a function of reflectance such that a given increase in the concentration of dye in the dyebath produces the same change in the value of the function, e.g. an increase from 0.1% to 0.2% must cause the same change as an increase from 1.0% to 1.1%. The function must, therefore, be linear. Additionally, this increase must also produce the same change if the reflectance has been reduced by the presence of any amount of any other dye: the function must, therefore, also be additive.

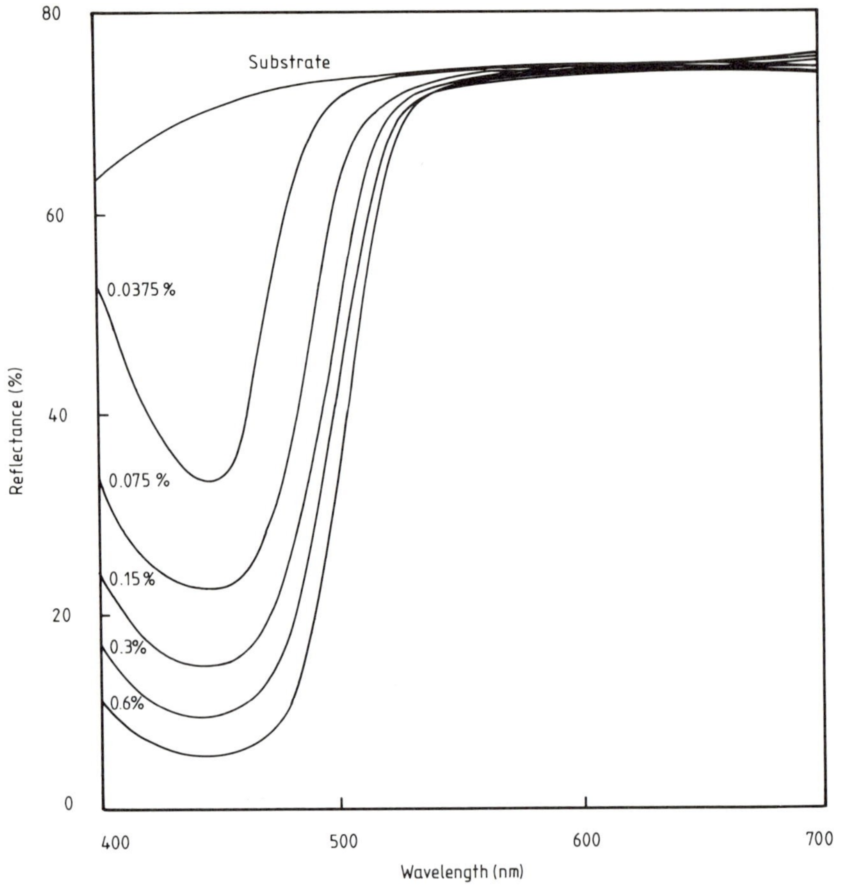

Figure 12.2 Reflectance curves of increasing depths of yellow dye.

In the case of solutions, we have already encountered such a function, the optical density, or the log of the reciprocal of transmittance when the latter is expressed on a 0–1 scale. In his early studies Atherton (1955) used a similar function of reflectance, $1/r$, because of its simplicity but in later work he and most other investigators used a more complex but much more reliable function. This had its origins in a study of radiation through a foggy atmosphere (Schuster 1905) from which Kubelka and Munk derived an exponential relationship between reflectance and a coefficient of absorption K and one of scatter S namely

$$r = \frac{S[1 - \exp(-2hQ)]}{K + S + Q - (K + S - Q)\exp(-2hQ)} \qquad Q = \sqrt{K(K + 2S)} \tag{12.1}$$

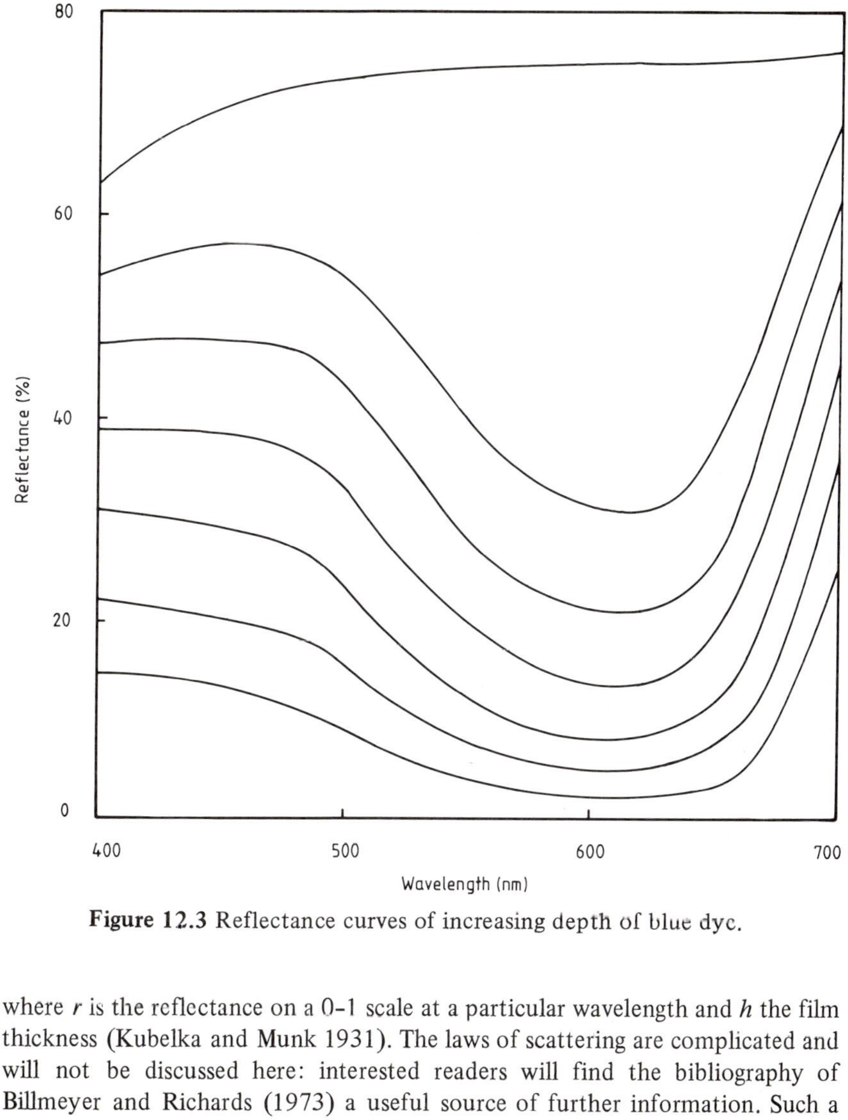

Figure 12.3 Reflectance curves of increasing depth of blue dye.

where r is the reflectance on a 0–1 scale at a particular wavelength and h the film thickness (Kubelka and Munk 1931). The laws of scattering are complicated and will not be discussed here: interested readers will find the bibliography of Billmeyer and Richards (1973) a useful source of further information. Such a relationship could, at that time, only be utilised graphically but later, in what Judd and Wyszecki (1975, p. 426) described as 'an amazing study', Kubelka derived equations involving the hyperbolic functions $\sin h$, $\cos h$ and $\cot h$ which could be solved numerically (Kubelka 1948). These coefficients are, to a first approximation, both linear and additive. If K and S are the values for unit concentrations of colorants, and K_W and S_W those for the substrate or white pigment present then the values for any combination of colorants 1, 2, 3 . . . present in concentrations $c_1, c_2, c_3 \ldots$ are as follows:

$$K_M = c_1 K_1 + c_2 K_2 + c_3 K_3 \ldots + K_W \qquad (12.2)$$

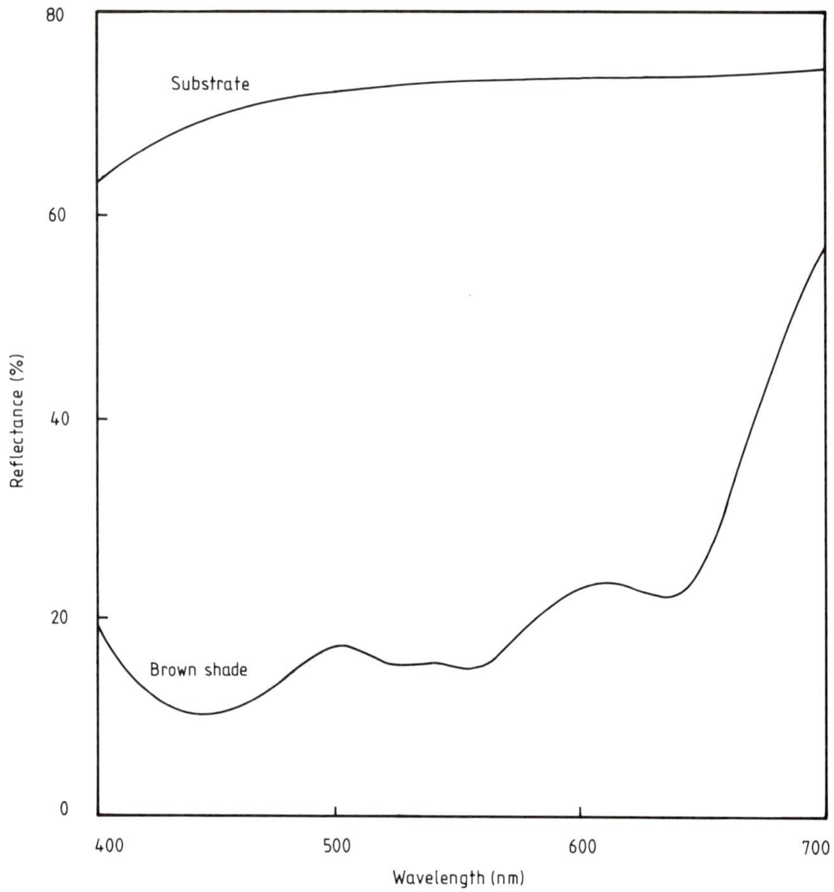

Figure 12.4 Reflectance curve of a brown dyeing made with red/yellow/blue mixture.

and

$$S_M = c_1 S_1 + c_2 S_2 + c_3 S_3 \ldots + S_W. \tag{12.3}$$

It was fortunate for the development of computer match prediction that the ratio K/S was related to reflectance in a surprisingly simple manner

$$K/S = (1 - r)^2/2r \tag{12.4}$$

and its inverse

$$r = K/S + 1 - [(K/S + 1)^2 - 1]^{0.5} \tag{12.5}$$

where r is the reflectance (on a 0–1 scale) of a specimen so thick that an increase in thickness has a negligible effect on the reflectance. This is termed 'reflectivity' and given the symbol r_∞. The subscript will be omitted for clarity but will always be implicit.

The function $(1-r)^2/2r$ was first used extensively for pigment identification with surface colours as the shape of the curve of log $(1-r)^2/2r$, like that of log density in the case of transmittance, is independent of colorant concentration. Pineo invented a cam for the Hardy which plotted log $(1-r)^2/2r$ as a function of wavelength (Pineo 1940), its use in pigment identification being described by Abbott and Stearns (1944) and as an aid in colour matching by Derby (1952). Its most important use, however, was for predicting the reflectance values of colorant mixtures from

$$(K/S)_M = \frac{c_1K_1 + c_2K_2 + c_2K_3 \ldots + K_W}{c_1S_1 + c_2S_2 + c_3S_3 \ldots + S_W} \tag{12.6}$$

and a further simplification arose from the fact that in the case of dyes, and of pigments used to produce pale shades, virtually all of the scatter is caused by the substrate or white pigment present. c_1S_1, c_2S_2 and $c_3S_3 \ldots$ can therefore be set to zero so that

$$\begin{aligned}(K/S)_M &= (c_1K_1 + c_2K_2 + c_3K_3 \ldots + K_W)/S_W \\ &= c_1(K_1/S_W) + c_2(K_2/S_W) + c_3(K_3/S_W) \ldots + K_W/S_W.\end{aligned} \tag{12.7}$$

This overall simplification made it possible for Davidson and Hemmendinger to market the first computer for match prediction in 1958. The computer was an analog computer.

1958–68, the era of the analog match prediction computer

The computer which Davidson and Hemmendinger designed is shown in figure 12.5 and a comprehensive account of the method of using it was given at a symposium organised by the Society of Dyers and Colourists in 1963 (Davidson *et al* 1963). It became affectionately known as COMIC (Colorant Mixture Computer) and the equation it solved was

$$\Delta(K/S) = (K/S)_S - c_1(K/S)_1 + c_2(K/S)_2 + c_3(K/S)_3 \ldots \tag{12.8}$$

the modification being the introduction of the term $(K/S)_S$ which is the K/S value of the target shade derived from equation (12.4) and $\Delta(K/S)$ which is the difference between $(K/S)_S$ and the predicted value. The value of $(K/S)_S$ was corrected for absorption by the substrate which eliminated the term $(K/S)_W$.

The numerical values of K/S at sixteen wavelengths (400, 420 ... 700 nm) for a known concentration of each dye are represented by the settings of sixteen resistances in individual plug-in boxes, up to five of which can be inserted into the computer for matching purposes. The K/S values of the target shade are similarly represented by resistances whose values are set by the sixteen dials at the top of the control panel. On the oscilloscope these values appear as sixteen dots representing an inverse of the reflectance curve – the lower the reflectance, the greater the absorption, the higher the value of K/S. The operator then chooses three suitable dyes and inserts the corresponding plug-in boxes. Adjust-

Figure 12.5 The first computer for match prediction.

ment of the first three concentration dials in the lower centre of the panel brings the dots down toward a horizontal reference line representing $\Delta(K/S) = 0$ and the operator's task is to get each dot on the line: the similarity between this task and the modern video game is obvious. If the operator could not achieve linearity he would try different three-dye combinations. If this failed then four- or five-dye combinations could be tried. This was, however, completely unsatisfactory as the use of more than three dyes or chromatic pigments is only acceptable under exceptional circumstances, i.e. when metamerism must be eliminated and this cannot be achieved with three-colorant mixtures.

At this stage the whole operation has been completely independent of the CIE system of colour measurement and as it was usually impossible to achieve linearity with any three-dye combination Davidson and Hemmendinger resorted to the CIE system by incorporating a tristimulus difference computer (TDC) into COMIC. This was based on the first adequate theoretical treatment of match prediction using the CIE system made by Park and Stearns (1944).

If a perfect match is obtained, whether metameric or not, the tristimulus values of the match will be identical with those of the target shade for a given illuminant/observer combination. ΔX, ΔY and ΔZ will therefore be zero and these values are given by

$$\Delta X = \sum S\bar{x}\Delta r \, \mathrm{d}\lambda\dagger \tag{12.9}$$

and the corresponding equations for ΔY and ΔZ, where Δr is the difference in reflectance between the match and the target shade at wavelength λ. As r can be calculated from K/S using equation (12.5), Δr can be obtained from $\Delta(K/S)$, which the computer has solved, using the relationship

$$r = \Delta(K/S) \times \mathrm{d}r/\mathrm{d}(K/S). \tag{12.10}$$

Substituting this value in equation (12.9) gives

$$X = \sum S\bar{x}\Delta(K/S)[\mathrm{d}r/\mathrm{d}(K/S)] \, \mathrm{d}\lambda \tag{12.11}$$

and similar values for ΔY and ΔZ. Davidson and Hemmendinger issued tables of values for $\mathrm{d}r/\mathrm{d}(K/S)$ as a function of r which were then set on the sixteen dials on the left of the console. The concentration dials were then adjusted until the three meters above the $\mathrm{d}r/\mathrm{d}(K/S)$ dials read zero.

When the first prediction was evaluated it was rarely close enough to be a good commercial match. The COMIC could be used to determine what corrections should be made either by measuring the reflectance values and entering the 16 K/S values into the computer via the dials below those used for the target shade or by determining ΔX, ΔY and ΔZ with a colorimeter and using the TDC: most operators, however, preferred to utilise their professional expertise and deduce the necessary adjustments from visual inspection.

About 200 COMIC were sold in the USA, UK, Western Europe, South Africa and Japan between 1958 and 1968 and a British analog computer of similar design was marketed by Redifon for a few years from 1965. The most interesting feature of this instrument was that it was interfaced to a spectrophotometer so that the sixteen potentionmeter dials were set automatically (Cutler 1965). Davidson and Hemmendinger had already interfaced a COMIC with a Hardy at the specific request of one of their American customers but had found the cost to be 'far beyond a reasonable level' (Tilleard 1964).

The use of the digital computer for match prediction

In 1955 Atherton of the Dyestuffs Division of ICI built an analog device for match prediction based on the shadows of wires bent to represent reflectance functions (Atherton 1955) and in the following year he wrote what was probably the first match prediction program for a digital computer. The computer was the Ferranti Pegasus and the program used K/S values as it was designed for textile dyeing (Alderson *et al* 1961). A single prediction took

† The CIE symbol for relative spectral power distribution is S: as this symbol has always been used for the Kubelka–Munk coefficient of scattering, care must be taken to avoid confusion.

20–60 seconds of computing time and although this was remarkably fast it will be interesting later to compare this time with that taken by a modern computer. In 1957, Duncan, whose first prediction had been made by a mixture of graphical and mechanical calculator methods, also wrote a program for the same computer using separate K and S values for paint matching (Duncan 1964).

Most of the programs for digital computation which have been written over the years have been designed to obtain identical tristimulus values from the start and not identical reflectance curves. The basic method of achieving this is to use the equation

$$K/S = c_1K_1 + c_2K_2 + c_3K_3 \qquad (12.12)$$

which is a simplified version of equation (12.7). From equation (12.5), r can be calculated and from the sixteen values of r, XYZ values can be obtained in the usual manner.

The XYZ values of the target shade can therefore be expressed by three non-linear equations in which the only unknowns are c_1, c_2 and c_3. It is possible in principle to solve these equations but only by a method of successive approximations. This is the method Park and Stearns used but it could take up to 2 hours (Stearns 1951). These equations can, however, be easily solved by a digital computer using the generalised Newton–Raphson iterative algorithm. Allen also devised an iterative solution based on an inverted matrix (Allen 1966). Both employ partial differentials which correspond to the $\mathrm{d}r/\mathrm{d}(K/S)$ familiar to COMIC users. They can be regarded as mathematical expressions of the way in which the concentrations of the dyes change with the position in XYZ space of the point representing the calculated match. Over small volumes of XYZ space these partial differentials are reasonably constant. They are then applied to the differences in X, Y and Z between the first solution and those of the target shade and Δc_1, Δc_2 and Δc_3 calculated. The revised values are then taken and the operation repeated until either the XYZ values are virtually identical, say within ± 0.05, or a pre-set number of iterations have been carried out at which point the attempt is abandoned.

In most respects the digital computer approach to match prediction was far better than that of the analog computer, the most important exception being cost. The Pegasus computer cost about £100 000 in 1960 whereas a COMIC + TDC + 100 plug-in boxes cost about one tenth of this sum. (Inflation has made £100 000 in 1960 equivalent to £750 000 in 1985.) Whilst it was possible to hire time on such computers which meant that a prediction could be obtained for about £1.50 at 1982 prices, this was not an attractive proposition until it became possible to install a terminal in the colour laboratory linked to the time-share computer by telephone and this did not occur in the UK until 1968. The potential market for analog match prediction computers therefore existed only as long as the much superior digital approach was either prohibitively expensive or operationally unattractive.

The next landmark in the history of match prediction was the introduction of Instrumental Match Prediction (IMP) by ICI Dyestuffs Division (Alderson *et al* 1963). This was based on a computer (Elliott 803B) installed solely for match prediction, i.e. a dedicated computer – and customers of the Division were encouraged to install colorimeters and to telex *RGB* values together with essential details concerning substrate, dyeing method and dye class. A computer match prediction would then be returned within an hour or so. This enabled even the smallest concern to experience the undeniable benefits of match prediction at the cost of only a few hundred pounds for a colorimeter. Even this expenditure was rendered unnecessary in 1970 with the introduction of the *ICI Colour Atlas* described in chapter 7: the *XYZ* values of the 1379 colour samples were stored in the computer which had also been programmed to calculate the effect of any one of the 25 grey filters used to obtain a match. Within a few months of the introduction of IMP the American Cyanamid Company introduced a similar service and during the following years all the large dye manufacturers installed computers for match prediction as part of their technical service though no others issued a general invitation to their customers to submit *RGB* or *XYZ* values.

In 1966 Pretema marketed the Pretema Colour Computer (PCC) which was a digital computer storing the *K*/*S* values of each dye on magnetic tape. As soon as the sixteen reflectance values of the target shade had been entered via the keyboard and the three wavelengths chosen – ideally those of the absorption peaks – the computer generated the first prediction which gave the same reflectance values at these wavelengths together with the *XYZ* values and those of the target shade. If these were not close enough the operator had to modify the prediction and input the new concentrations until they were (Rohner 1965). This operation was actually more difficult than the corresponding stage with COMIC and it is surprising that the computer was not programmed to carry out iteration as methods of doing this had already been published.

In 1966 IBM supplied a match prediction program for users of their 1130 computer which, needless to say, was fully capable of carrying out all the calculations involved in match prediction. In 1967 a digital version of the COMIC was marketed as COMIC II but, unfortunately, the computer specially built for this purpose was inadequate and manufacture ceased shortly afterwards. Before considering the next and the most important stage in the development of computer match prediction it is convenient to refer to the major modifications which have been made in the mathematics of match prediction most of which were only possible with digital computers.

The use of separate K *and* S *values*

It is always better to use separate *K* and *S* values for match prediction in the case of paints and of plastics but absolute values are not necessary: relative values will suffice and the following method of determining them is an extension of that devised by Duncan (1949).

A white pigment is chosen as standard and given as S_W value of 1: K_W is then the *K*/*S* value of a paint made with 100% of this white pigment. Paints are then made up with a coloured pigment alone (mass tone) and with this pigment let down with the white pigment such that $c+w=1$ where c and w are the concentrations of the coloured pigment and the white pigment. Reflectance curves are then plotted and K/S values calculated

$$(K/S)_{W} \quad \text{(white)}$$

$$(K/S)_{MT} \quad \text{(mass tone)}$$

$$(K/S)_{LD} \quad \text{(let down).}$$

The coefficient of scatter, S, of the coloured pigment is then given by

$$S = \frac{\text{parts of white}}{\text{parts of colour}} \times \frac{(K/S)_{LD} - K_{W}}{(K/S)_{MT} - (K/S)_{LD}}$$

and

$$K = S \times (K/S)_{MT}.$$

If the reflectance of a mass tone of the coloured pigment is as high at some wavelengths as the white pigment then the coloured pigment (CP) should be mixed with a black pigment (B) when

$$S = \frac{S_{B}}{\text{Ratio CP:B}} \times \frac{(K/S)_{\text{MT black}} - (K/S)_{\text{mixture}}}{(K/S)_{\text{mixture}} - (K/S)_{\text{MT yellow}}}$$

and

$$K = S \times (K/S)_{\text{MT yellow}}.$$

The Saunderson correction for paints and plastics

This was devised (Saunderson 1942) to take into account specular reflectance and internal reflectance which are not allowed for in basic Kubelka–Munk theory.

The Saunderson-corrected reflectance, r', is

$$r' = \frac{r - k_1}{1 - k_1 - k_2(1 - r)}$$

where k_1 is the Fresnel reflection coefficient we encountered in chapter 9. This can usually be set at 0.04 and k_2 is an empirical constant for which the value of 0.4 suggested by Saunderson has proved satisfactory. Because spectrophotometers vary in the extent to which they exclude the specular component it is better, for match prediction purposes, to measure with the specular component included and to apply the Saunderson correction.

The Pineo correction for textiles

This was devised to allow for the fact that the deepest dyed textile does not have zero reflectance. It is

$$K/S = [1 - (r - r_0)]^2/2(r - r_0)^{1/n}$$

where r_0 is the minimum reflectance and n an arbitrary constant for which experience has shown the best value to be 1 (Pineo 1940). The best method of determining r_0 is not, as one might expect, to dye a full black shade but to dye increasing depths of a yellow: the minimum reflectance achieved is invariably less than that of any black and is used as r_0 for all wavelengths.

Non-linear K/S *values*

Both the Pineo and the Saunderson corrections improve the linearity of K and S with concentration but there is often a significant residual non-linearity with textile dyes after applying the Pineo correction. This is mainly due to the fact that the dye on the fibre at the end of the dyeing operation is usually not linearly related to that in the dyebath at the start. McDonald *et al* (1976) allowed for this by using the thermodynamics of the dyeing process to predict the amount of dye in the dyebath required to give the amount on the fibre which the Kubelka-Munk approach gives. The laboratory analyses required, however, are so extensive that this method has rarely been used and the most satisfactory method is a simple two-stage prediction. A straight line is fitted to that portion of the K/S against concentration curve which is linear – at low values – and the corresponding K/S value used to derive the initial prediction. Using a conventional non-linear interpolation formula, the true K/S values for the concentrations predicted are calculated and used in the final prediction.

Metamerism

Because the digital computer generates a sixteen-point reflectance curve, if the corresponding values for the target shade are also known, the CIE index of metamerism can be calculated for any illuminant. Whilst this can be restricted to those sources under which the goods may be inspected a better method is to calculate ΔE under Philips Colour 84 and the Westinghouse metamer detecting lamp as described in chapter 10.

The era of systems marketing, 1969 . . .

By 1969, in spite of the availability of digital computers and programs, the growth rate of companies using match prediction was only rising very slowly. The chief reason for this state of affairs was that two separate pieces of expensive equipment – the spectrophotometer and the computer – had to be purchased from different companies and a program obtained or written.

> To get these functioning properly for routine match prediction an expert is needed who knows how to work with colours, who is fully acquainted with the conditions and requirements of the dyehouse, who has a sufficient knowledge of colour measurement and computing recipes, who is able to assist in programming and who is capable of supervising the subsequent routine operation of the computer installation. Alternatively, a team capable of dealing with these requirements must be assembled.

(Thurner and Würz 1966). These obstacles were formidable to all except the largest companies and only began to be overcome with the introduction of the concept of complete systems marketing by two companies: Instrumental Colour Systems Limited in the UK in 1969 and Applied Color Systems Incorporated in the USA in 1970. This development was, in one respect, a natural progression once it became feasible to interface a spectrophotometer to a digital computer because this could be powerful enough to carry out the calculations necessary for match prediction and to have adequate capacity to store the reflectance values for all the calibration dyeings or pigment formulations. If the electronic engineering expertise in such a company was augmented by the kind of experts Thurner and Würz described, a single company could design, assemble, program and install a complete match prediction unit, train staff for routine operation and program modification if desired and provide a continuous maintenance service for every component. The experts are, of course, very much more cost effective in a systems company as their expertise benefits hundreds of companies, not one. Many of the large companies who pioneered match prediction in the 1960s have found it better to replace their obsolete units with those developed by systems companies of which four were prominent in 1982.

The original purpose of computer match prediction was to eliminate the lengthy and costly stage of producing in the laboratory a dye recipe or pigment formulation which would match the shade in question. This benefit, however, has now become of secondary importance as the speed of modern computers has made automatic colorant selection feasible with much greater cost savings. Though this technique will be described in a textile dyehouse context, it is equally applicable to any coloration process where there is more than one way of achieving the desired shade.

Automatic dye selection

For a given fibre and specific fastness properties a dyehouse will normally hold in stock a minimum of ten to twelve chemically different dyes from various manufacturers. Over the years the dyer has built up a shade bank containing recipes for hundreds or even thousands of shades, no recipe normally containing more than three dyes. These recipes have withstood the test of time but one factor was rarely, if ever, considered before the introduction of computer match prediction: the cost of each in comparison with other three-dye combinations which would have been equally suitable technically.

The speed of modern computers is now so great that, instead of the dyer identifying the three dyes to be used, predictions can be made of every possible three-dye combination (or four- or five-dye combination if needed for completely eliminating metamerism). At this stage the prudent dyer will increase the number of potentially useful dyes as up to twenty can be handled by the computer in a reasonable time. Every possible three-dye combination will be tried – there is no point in trying to program the computer to eliminate attempts to match a red shade with two yellows and a blue, for example.

Successful combinations are then printed out in a form which will permit the dyer to make the best choice. This will not always be the cheapest because its metameric index may be too great but by printing out combinations in order of increasing cost within ranges of metameric indexes the right choice can be made. When this has been done for a large number of shades it is usually found that some dyes in stock never appear in the chosen predictions and there can then be further savings by reducing the inventory.

Minimum pigment loading

In the areas of surface coatings even greater cost savings are possible by using Kubelka-Munk calculations to determine the minimum pigment loading required to give the preferred hiding power as assessed by coating a black and white card with a coating of the appropriate thickness (Best 1981). The same technique is also applicable to achieving the desired opacity in a plastic with minimum pigment loading.

Extensions to other users

Modern minicomputers, which are ideal for match prediction, can be accessed independently by other users, each with his own spectrophotometer – for prediction, recipe correction or quality control etc – or merely a terminal if the computer is to be used for other purposes. Amongst the infinite variety of other purposes the production of complete work sheets, stock control with re-ordering signals, control of semiautomatic weighing machines and automatic dyeing machines, may be mentioned.

Figure 12.6 A typical modern computer match prediction system.

A typical modern system for match prediction by a single user is shown in figure 12.6. Such a system would cost less than £40 000 (at 1982 prices) and, whereas at the start of digital computer match prediction, computing a recipe having identified the three colorants to be used would take between 20 and 60 seconds, today the time should not exceed 0.5 seconds.

Match prediction systems are now operating successfully in textile dye-houses and printworks, and in the paint, plastics, paper (dyeing and printing), cosmetics, food and candle manufacturing industries (Best 1981) saving millions of pounds annually in colorant costs alone. Experience has shown that the installation of a modern match prediction system has resulted in savings in dye costs of between 10 and 32%. When three-band lamps were introduced (as described in chapter 8) extremely serious problems of metamerism occurred, some of which were virtually insoluble by traditional methods as four-dye combinations were essential. Match prediction programs invariably calculated the colour difference of every daylight match under tungsten light as well and it was an easy matter to extend this to include a typical three-band lamp especially if access to the program by the user was permitted. Park and Thompson (1982) found that this facility was 'probably the biggest timesaver'. This, is, unquestionably, the most impressive achievement of the method of colour measurement envisaged by Maxwell in the middle of the last century and brought to fruition by the pioneers who, in 1931, created the CIE system.

References

Abbott R and Stearns E I 1944 *American Cyanamid Co. Tech. Bull.* 754

Abney W de W 1913 *Researches in Colour Vision and the Trichromatic Theory* (London)

Acree W L 1983 *AATCC–ISCC Symposium on Practical Applications of Color Control* p. 32

Adams E Q 1923 *Psych. Rev.* **30** 56

— 1942 *J. Opt. Soc. Am.* **32** 168

AIC 1983 *The Forsius Symposium on Colour Order Systems* (Stockholm: The Scandinavian Colour Institute)

Alderson J V, Atherton E and Derbyshire A N 1961 *J. Soc. Dye. Col.* **77** 657

Alderson J V, Atherton E, Preston C and Tough D 1963 *J. Soc. Dye. Col.* **79** 723

Allen E 1966 *J. Opt. Soc. Am.* **56** 1256

Allen N S and McKeller J F 1980 *Photochemistry of Dyed and Pigmented Polymers* (Barking, England: Applied Science) p. 266

Alman D H and Billmeyer F W Jr 1976 *Col. Res. Appl.* **1** 141

Arnold S J, Ogryzlo E A and Witzke H 1964 *J. Chem. Phys.* **40** 1769

Atherton E 1955 *J. Soc. Dye. Col.* **71** 389

Bailey K C 1929 *The Elder Pliny's Chapters* (London: Arnold)

Baker J T 1974 *Endeavour* **34** 11

Baker J T and Duke C C 1973 *Aust. J. Chem.* **26** 2153

Baker J T and Sutherland M D 1968 *Tetrahedron Lett.* 43

Ball S, Goodwin T W and Morton R A 1946 *Biochem. J.* **40** 59

Bamford C H 1949 *J. Soc. Dye. Col.* **65** 680

Bamford C H and Dewar M G S 1949 *J. Soc. Dye. Col.* **65** 674

Barlow H B 1982 *Vis. Res.* **22** 635

Bartleson C J 1976 *Col. Res. Appl.* **1** 181

— 1977 in *Color 77* (Bristol: Adam Hilger) p. 64

Beatty J M 1961 Trade Winds *Saturday Review (USA)* Aug 5 p. 6

Beer M and Longuet-Higgins H C 1955 *J. Chem. Phys.* **23** 1390

Bello F 1959 *Fortune* **59** (5) 144

Bentley P, McKellar J F and Phillips G O 1974 *Rev. Prog. Col. Rel. Topics (Soc. Dye. Col.)* **5** 33

Berger A and Brockes A 1966 in *Tag. Ber. Intern. Farbtagung Luzern, 1965* (Göttingen: Musterschmidt) p. 331

Berger A and Strocka D 1973 *Appl. Opt.* **12** 338
Berlin B and Kay P 1969 *Basic Color Terms: Their Universality and Evolution* (Berkeley and Los Angeles: University of California Press)
Berns R S and Billmeyer F W Jr 1985 *Col. Res. Appl.* **10** 73
Best R P 1981 in *Golden Jubilee of Colour in the CIE* (Bradford: Society of Dyers and Colourists) p. 139
Billmeyer F W Jr 1983 *Col. Res. Appl.* **8** 192
Billmeyer F W Jr and Alessi P J 1981 *Col. Res. Appl.* **6** 195
Billmeyer F W Jr and Hemmendinger H 1981 in *Golden Jubilee of Colour in the CIE* (Bradford: Society of Dyers and Colourists) p. 98
Billmeyer F W Jr and Rich D C 1978 *Plastics Eng.* **34** 12 p. 35
Billmeyer F W Jr and Richards L W 1973 *J. Col. App.* **2** (No. 2) 4
Billmeyer F W Jr and Saltzman M 1981 *Principles of Color Technology* 2nd edn (New York: John Wiley)
Bird C L 1945 *J. Soc. Dye. Col.* **61** 321
Birren F 1979 *Col. Res. Appl.* **4** 171
Bitensky M W, Gorman R E and Miller 1971 *Proc. Natl. Acad. Sci.* **68** 561
Blakey R R and Landon G 1978 *Measuring Colour* (Billingham: BTP Tioxide Ltd)
Blough D S and Schrier A M 1963 *Science* **139** 493
Bolland J L and Cooper H R 1954 *Proc. R. Soc.* A **225** 405
Bowen E J 1949 *J. Soc. Dye. Col.* **65** 613
Bowmaker J K and Dartnall H J A 1980 *J. Physiol.* **298** 501
Bownds D 1967 *Nature* **216** 1178
Brewster D 1834 *Trans. R. Soc. Edin.* **12** 132
Bridge N K 1960a *J. Soc. Dye. Col.* **76** 484
— 1960b *Trans. Farad. Soc.* **56** 1001
Bridge N K and Porter G 1958 *Proc. R. Soc.* A **244** 259, 275
Bridges C D B 1965 *Vision Res.* **5** 223
— 1967 *Vision Res.* **7** 349
Brockes A 1965 in *Tag. Ber. Intern. Farbtagung Luzern* (Göttingen: Musterschmidt) p. 811
— 1970a *Die Farbe* **19** 135
— 1970b in *Color 69* (Göttingen: Musterschmidt) p. 752
— 1975 *Textilveredlung* **10** (2) 47
Brown P K and Wald G 1964 *Science* **144** 45
Brunello F (transl. Hickey B) 1973 *The Art of Dyeing in the History of Mankind* (Vicenza: Neri Pozza Editore)
Burawoy A 1939 *J. Chem. Soc.* 1177
Burdett B C 1982 *J. Soc. Dye. Col.* **98** 114
Burkamp W 1923 *Z. Sinnesphysiol.* **55** 307
Busch G E, Applebury M L, Lamola A A and Rentzepis P M 1972 *Proc. Natl. Acad. Sci.* **69** 2802
Calvert J G and Pitts J N 1966 *Photochemistry* (New York: Wiley)
Carbone J 1962 *Proc. Can. Text. Seminar* p. 85
Caro H 1892 *Ber. Deut. Chem. Ges.* **25** 955
Cervetto L and Fuortes M G F 1978 *Ann. Rev. Biophys. Bioeng.* **7** 229
Chickering K D 1967 *J. Opt. Soc. Am.* **57** 537
Christie J S and McConnell G 1977 in *Color 77* (Bristol: Adam Hilger) p. 309

Christopherson C, Wätjen F, Buchardt O and Antoni U 1978 *Tetrahedron* **34** 1779

CIE 1971 *Special Metamerism Index: Change in Illuminant* (Paris: CIE)

— 1974 *Method of Measuring and Specifying Colour Rendering Properties of Light Sources* (Paris: CIE)

— 1978 *Recommendations on Uniform Colour Spaces – Colour-Difference Equations – Psychometric Colour Terms* (Paris: CIE)

— 1981 *A Method for Assessing the Quality of Daylight Simulators for Colorimetry* (Paris: CIE)

Clarke F J J 1980 *CIE Proc., 19th Session* (Paris: CIE)

Clarke F J J, McDonald R and Rigg B 1984 *J. Soc. Dye. Col.* **100** 128

Clarke F J J and McKinnon R A 1982 *Col. Res. Appl.* **7** 257

Clarke F J J and Samways P R 1968 *NPL Report* MC2

Coates E 1967 *J. Soc. Dye. Col.* **83** 95

Coates E, Day S, Provost J R and Rigg B 1972 *J. Soc. Dye. Col.* **88** 186

Cole B L, Henry G H and Nathan J 1966 *Vision Res.* **6** 301

Cole W 1685 *Phil. Trans. R. Soc.* **15** 1278

Colour Measurement Committee 1976 *J. Soc. Dye. Col.* **92** 407

Committee on Colorimetry 1953 *The Science of Color* (New York: Crowell)

Compton J A 1984 *Col. Res. Appl.* **9** 15

Cooper A C 1948 *J. Oil Col. Chem. Assoc.* **31** 343

— 1975 paper presented at SDC Symposium, Nottingham

Cooper A C and McLaren K 1973a *J. Soc. Dye. Col.* **89** 41

— 1973b *J. Oil Col. Chem. Assoc.* **56** 134

Corey R J and Taylor W C 1964 *J. Am. Chem. Soc.* **86** 3881

Cornwell-Clyne A 1951 *Colour Cinematography* (London: Chapman and Hall)

Cunliffe P W 1930 *J. Soc. Dye. Col.* **46** 108

Cutler A E 1965 *J. Soc. Dye. Col.* **81** 601

Dahne S and Kulpe S 1978 *J. Prakt. Chem.* **320** 395

Dandridge A G, Drescher H A E and Thomas J 1929 *British Patent* 322169

Dartnall H J A 1953 *Brit. Med. Bull.* **9** 24

Dartnall H J A, Bowmaker J K and Mollon J D 1983 *Proc. R. Soc.* B **220** 115

Dartnall H J A and Lythgoe J N 1965 *Vision Res.* **5** 81

Davidson H R and Friede E 1953 *J. Opt. Soc. Am.* **43** 581

Davidson H R, Hemmendinger H and Landry J L R 1963 *J. Soc. Dye. Col.* **79** 577

Davidson H R and Imm L W 1949 *J. Opt. Soc. Am.* **39** 942

Davies A K, Ford R, Gee G A, McKellar J F and Phillips G O 1972 *Chem. Commun.* 873

Davies A K, Gee G A, McKellar J F and Phillips G O 1973a *J. Chem. Soc. Perkin Trans.* **2** 1742

— 1973b *Chem. Ind.* 431

Dearman H H and Chan A 1965 *J. Chem. Phys.* **44** 416

Dent C E, Linstead R P and Lowe A R 1934 *J. Chem. Soc.* 1036

Derby R E Jr 1952 *Am. Dyes. Rep.* **41** 550

Derbyshire A N and Marshall W J 1980 *J. Soc. Dye. Col.* **96** 166

de Diesbach H and von der Weid E 1927 *Helv. Chim. Acta* **10** 886

Dimmick F L and Hubbard M R 1939 *Am. J. Psychol.* **52** 242, 348

Duncan D R 1949 *J. Oil Col. Chem. Assoc.* **32** 296
— 1963 *J. Soc. Dye. Col.* **79** 603
— 1964 *Paint Research Station Bull.* **13** 132
Eastwood D 1973 in *Colour 73* (London: Adam Hilger) p. 293
Egerton G S 1947 *J. Soc. Dye. Col.* **63** 161
— 1948 *J. Text. Inst.* **39** T305
— 1949 *J. Soc. Dye. Col.* **65** 764
Egerton G S and Morgan A G 1970 *J. Soc. Dye. Col.* **86** 242
— 1971 *J. Soc. Dye. Col.* **87** 268
Ermolaev V L 1963 *Sov. Phys.-Usp.* **80** 333
Estévez O 1979 On the Fundamental Data-base of Normal and Dichromatic Colour Vision *PhD Thesis* University of Amsterdam (Meppel: Krips Repro)
Fabian J and Hartmann H 1980 *Light Absorption of Organic Colorants* (Berlin: Springer)
Fastness Tests Co-ordinating Committee 1953 *J. Soc. Dye. Col.* **69** 404
— 1983 *J. Soc. Dye. Col.* **99** 101
Fletcher R and Voke J 1985 *Defective Colour Vision – fundamentals, diagnosis and management* (Bristol: Adam Hilger)
Foote C S and Wexler S 1964 *J. Am. Chem. Soc.* **86** 3879, 3880
Fouquet H and Bielig H-J 1971 *Angew. Chemie. Int. Edn.* **10** 816
Friedländer P 1906 *Liebig's Ann.* **351** 390
— 1909 *Z. Ang. Chem.* 2321
Friele L F C 1961 *Farbe* **10** 193
— 1978 *Col. Res. Appl.* **3** 53
Fries W 1981 *Proc. R. Soc.* B **213** 73
Fukuda Y, Kawamura S, Yoshizawa T and Miki N 1981 *Biochim. Biophys. Acta* **675** 188
Fung B K-K, Hurley J B and Stryer L 1981 *Proc. Natl. Acad. Sci.* **78** 152
Gall L 1973 in *Colour 73* (London: Adam Hilger) p. 153
Gall L and Riedel G 1965 *Farbe u. Lack* **71** 891
Garston B 1980 *J. Soc. Dye. Col.* **96** 535
Giles C, Hojiwala B J and Shah C D 1972 *J. Soc. Dye. Col.* **88** 403
Glasser L G, McKinney A H, Reilly C D and Schnelle P D 1958 *J. Opt. Soc. Am.* **48** 736
Glasser L G and Troy D J 1952 *J. Opt. Soc. Am.* **42** 652
Godlove I H 1951 *J. Opt. Soc. Am.* **41** 760
Gogala M, Hamdorf K and Schwemer J 1970 *Zeit. vergl. Physiol.* **70** 410
Gouras P and Zrenner E 1981 *Prog. Sens. Physiol.* **1** 139
— 1982 *Col. Res. Appl.* **7** 205
Grassman H G 1853 *Poggendorf's Ann.* **89** 69
— 1854 (transl. anon) *Phil. Mag.* **4** 254
Green B, Monger T, Alfano R, Aton B and Callender R 1977 *Nature* **264** 179
Griffiths J 1972 *Chem. Soc. Rev.* **1** 481
— 1976 *Colour and Constitution of Organic Molecules* (London: Academic)
— 1984 in *Review of Progress in Coloration* (Bradford: Society of Dyers and Colourists) p. 21
Griffiths J and Hawkins C 1973 *J. Soc. Dye. Col.* **89** 173
— 1976 *Polymer* **17** 1113
Grundfest H 1958 *Arch. Ital. Biol.* **96** 135

Guild J 1931 *Phil. Trans. R. Soc.* Ser. A **230** 149
— 1932 in *Some Problems of Visual Perception* (London: Physical Society)
Gundlach D 1977 in *Color 77* (Bristol: Adam Hilger) p. 218
Hagins W A 1972 *Ann. Rev. Biophys. Bioeng.* **1** 144
Haller R 1924 *Melliand Textilber.* **5** 543
Hård A and Sivik L 1981 *Col. Res. Appl.* **6** 129
Hardy A C 1929 *J. Opt. Soc. Am.* **18** 96
— 1935 *J. Opt. Soc. Am.* **25** 305
Hardy A C and Pineo O W 1932 *J. Opt. Soc. Am.* **22** 430
Hargrave P A, McDowell J H, Curtis D R, Wang J K, Juszczak E, Fong S-L, Mohana Rao J K and Argos P 1984a *Biophys. Str. Mech.*
Hargrave P A, McDowell J H, Fieldman R J, Atkinson P H, Rao J K M and Argos P 1984b *Vis. Res.* **24** 1487
Harry S in *Color 69* (Göttingen: Musterschmidt) p. 1090
von Helmholtz 1852 *Ann. Phys., Lpz.* **9** 524 (Engl. transl. 1852 *Phil. Mag.* **4**)
— 1855 *Ann. Phys., Lpz.* **94** 1
— 1866 *Handbuch der Physiologischen Optik* (Hamburg: Voss)
— 1924 *Physiological Optics* transl. Southall P C (New York: Optical Society of America)
Hemmendinger H 1972 *J. Col. Appl.* **1** No. 6 11
Hering E 1878 *Zur Lehre vom Lichtsinne* (Vienna: Carl Gerold's Sohn)
— 1964 *Outlines of a Theory of the Light Sense* transl. Hurvich L M and Jameson D (Cambridge, Mass.: Harvard)
Holbro T 1962 *Symposium on Recent Advances in Colouring Matters* (College of Science and Technology, Manchester)
Honig B and Dinur U 1979 *J. Am. Chem. Soc.* **101:23** 7084
Hubbard R and Kropf A 1958 *Proc. Natl. Acad. Sci. USA* **44** 130
Hubel D H and Wiesel T N 1968 *J. Physiol.* **195** 215
— 1979 *Sci. Am.* **241** 150
Hunt R W G 1967 *The Reproduction of Colour* 2nd edn. (London: Fountain Press) p. 233
— 1982 *Col. Res. Appl.* **7** 95
Hunt R W G and Pointer M R 1985 *Col. Res. Appl.* in the press
Hunter R S 1940 *J. Opt. Soc. Am.* **30** 536
— 1942 *J. Opt. Soc. Am.* **32** 509
— 1948 *J. Opt. Soc. Am.* **38** 1094
Hurvich L M and Jameson D 1955 *J. Opt. Soc. Am.* **45** 602
— 1960 *Ann. Rev. Psych.* **11** 99
ICI 1979 *Reactive Dyes* (London: Imperial Chemical Industries Limited)
Ingle D J 1985 *Science* **227** 651
Inhoffen H H 1950 *Liebig's Ann.* **570** 54
Ishihara S 1973 *Test for Colour-Blindness* (Tokyo: Kanehara Shuppan)
Isler O 1956 *Helv. Chim. Acta* **39** 249
ISO 1982 *International Standard 105* (Geneva: ISO)
Jacobs G H 1984 *Vis. Res.* **24** 123
Jaeckel S M 1973 *Appl. Opt.* **12** 1299
Judd D B 1930 *J. Res. NBS* **4** 515
— 1935 *J. Opt. Soc. Am.* **25** 24
— 1940 *J. Opt. Soc. Am.* **30** 2

— 1945 *J. Opt. Soc. Am.* **35** 199
— 1949 *J. Opt. Soc. Am.* **39** 945
— 1950 *J. Opt. Soc. Am.* **40** 52
— 1959 *J. Opt. Soc. Am.* **49** 322
— 1960 *J. Opt. Soc. Am.* **50** 254
Judd D B, MacAdam D L and Wyszecki G 1964 *J. Opt. Soc. Am.* **54** 1031
Judd D B and Wyszecki G 1975 *Color in Business, Science and Industry* (New York: Wiley)
Kahn A U and Kasha M 1963 *J. Chem. Phys.* **39** 2105
— 1964a *J. Chem. Phys.* **40** 605
— 1964b *Nature* **204** 241
Kaiser P K and Hemmendinger H 1980 *Col. Res. Appl.* **5** 65
Kaneko A 1970 *J. Physiol.* **207** 623
Karp A 1959 *Nature* **184** 710
Kasha M 1960 *Radiat. Res. Supp.* **2** 243
Katz D and Révész 1921 *Z. Angew. Psychol.* **18** 307
Kautsky H 1938 in *Luminescence,A General Discussion of the Faraday Society* (London: Gurney and Jackson) p. 216
Kearns D R 1971 *Chem. Rev.* **71** 395
Kelly K L 1981 *Col. Res. Appl.* **6** 59
Kiel E G and Heertjes P M 1965 *J. Soc. Dye. Col.* **81** 98
Kishner S J 1977 in *Colour 77* (Bristol: Adam Hilger) p. 305
Klessinger 1978 *Chem. in Unsere Zeit* **12** 1
König A and Dieterici C 1892 *Z. Psychol. Physiol. Sinnesorgane* **4** 231
Krais P 1929 *Melliand Textilber.* **10** 468
von Kries J 1905 in *Nagel's Handbuch der Physiologie des Menschen* vol. 3 (Braunschweig: Vieweg)
Krüger J and Fischer B 1983 in *Colour Vision* (London: Academic Press) p. 291
Krüger J and Gouras P 1980 *J. Neurophysiol.* **43** 1055
Kubelka P 1948 *J. Opt. Soc. Am.* **38** 448
Kubelka P and Munk F 1931 *Z. Tech. Phys.* **12** 593
Kuehni R G 1971a *J. Col. App.* **1** (1) 30
— 1971b *Text. Chem. Col.* **3** 248
— 1972 *J. Col. App.* **1** (3) 4
— 1975 *Computer Colorant Formulation* (Lexington, Mass.: D C Heath)
— 1977 *Col. Res. Appl.* **2** 187
— 1978 *Text. Chem. Col.* **10** 75
— 1983 *Col. Res. Appl.* **8** 192
— 1984 *J. Soc. Dye. Col.* **100** 281
Kuehni R G and Marcus R T 1979 *Col. Res. Appl.* **4** 83
de Lacaze-Duthiers H 1859 *Ann. Sci. Nat. (Zool.)* **12** 5
Ladd J H and Pinney J E 1955 *Proc. Inst. Radio Eng.* **43** 1137
Land E H 1959a *Proc. Natl. Acad. Sci.* **45** 115
— 1959b *Proc. Natl. Acad. Sci.* **45** 636
— 1959c *Sci. Am.* **200** (5) 84
— 1959d *Sci. Am.* **201** (1) 17
— 1964 *Am. Sci.* **52** 1964
— 1977 *Sci. Am.* **237** (6) 108
— 1983 *Proc. Natl. Acad. Sci.* **80** 5163
Land E H, Hubel D H, Livingstone M S and Burns M M 1983 *Nature* **303** 616

Land E H and McCann J J 1971 *J. Opt. Soc. Am.* **61** 1

Leaver I H 1980 in *Photochemistry of Dyed and Pigmented Polymers* ed. N S Allen and J K McKellar (Barking, England: Applied Science) chap. 4

Leroi-Gourhan A 1982 *Sci. Am.* **246** (12) 80

Lewis G N and Kasha M 1944 *J. Am. Chem. Soc.* **66** 2100

Linstead R P and Lowe A R 1944 *J. Chem. Soc.* 1022

Livingstone M S and Hubel D H 1984 *J. Neurosci.* **4** 309

Livingstone R 1949 *J. Soc. Dye. Col.* **65** 781

Lozano R D 1977 *Col. Res. Appl.* **2** 13

MacAdam D L 1934 *J. Opt. Soc. Am.* **24** 188

— 1937 *J. Opt. Soc. Am.* **27** 294

— 1942 *J. Opt. Soc. Am.* **32** 247

— 1943 *J. Opt. Soc. Am.* **33** 18, 675

— 1966 *J. Opt. Soc. Am.* **56** 1784

— 1969 *Col. Eng.* Nov–Dec 40

McCamy C S 1960 *J. Opt. Soc. Am.* **50** 510

McCann J J, McKee S P and Taylor T H 1976 *Vision Res.* **16** 445

McCann J J and Houston K L 1983 in *Colour Vision* (London: Academic Press) p. 535

McDonald R 1974 *J. Soc. Dye. Col.* **90** 189

— 1980a *J. Soc. Dye. Col.* **96** 372

— 1980b *J. Soc. Dye. Col.* **96** 486

— 1980c *J. Soc. Dye. Col.* **96** 418

— 1982 *J. Oil Col. Chem. Assoc.* **65** 43, 93

McDonald R, McKay D and Weedall P J 1976 *J. Soc. Dye. Col.* **92** 39

McLaren K 1956 *J. Soc. Dye. Col.* **72** 86

— 1966 *J. Soc. Dye. Col.* **82** 345, 382

— 1969 *Col. Eng.* **7** 38

— 1970a in *Color 69* (Göttingen: Munsterschmidt) p. 688

— 1970b *J. Soc. Dye. Col.* **86** 138

— 1971a *J. Oil Col. Chem. Assoc.* **54** 879

— 1971b in *Color Metrics* (Soesterberg, Holland. AIC) p. 296

— 1972 *J. Col. Appl.* **I** (4) 12

— 1973a *J. Col. Appl.* **II** (1) 57

— 1973b *Text. Chem. Col.* **9** 160

— 1975 *Melliand Textilber.* **56** 237

— 1976a *J. Soc. Dye. Col.* **92** 317

— 1976b *J. Soc. Dye. Col.* **92** 364

— 1978a *J. Soc. Dye. Col.* **94** 414

— 1978b *Col. Res. Appl.* **3** 155

— 1980 *Col. Res. Appl.* **5** 139

— 1981 in *Golden Jubilee of Colour in the CIE* (Bradford: Society of Dyers and Colourists) p. 168

— 1982 *J. Soc. Dye. Col.* **98** 256

— 1983a in *The Forsius Symposium on Colour Order Systems* (Stockholm: The Scandinavian Colour Institute) p. 17.1 (5)

— 1983b in *CIE 20th Session* Vol. 2 D109

— 1985a *Col. Res. Appl.* **10** in the press

— 1985b *Col. Res. Appl.* **10** 124

McLaren K and Perry A C 1979 *J. Soc. Dye. Col.* **95** 115

McLaren K and Taylor P F 1981 *Col. Res. Appl.* **6** 75
MacNichol E F Jr and Svaetichin G 1958 *Am. J. Ophthalmol.* **46** 26
Marcus R T 1978 *Col. Res. Appl.* **3** 29
Marks W B, Dobelle W H and MacNichol E F Jr 1964 *Science* **143** 1181
Marr D 1974 *Vis. Res.* **14** 1377
— 1982 *Vision* (San Francisco: Freeman)
Mason S F 1970 in *Chemistry of the Synthetic Dyes* vol. 3 ed. K Venkataraman (London: Academic)
Maxwell J C 1856 *Trans. R. Scot. Soc. Arts* **4** 394
— 1860 *Phil. Trans. R. Soc.* **150** 57
— 1890 *The Scientific Papers of James Clerk Maxwell* (London: Cambridge University Press)
Meier A 1971 in *The Chemistry of Synthetic Dyes* vol 4 ed. K Venkataraman (New York: Academic) chap. VII
Michael C R 1978 *J. Neurophysiol.* **41** 572, 1233, 1250
— 1981 *J. Neurophysiol.* **46** 587
Millich F and Oster G 1959 *J. Am. Chem. Soc.* **81** 1357
Mollon J D 1985a *The Listener* **113** (2891) 6
— 1986 *J. Soc. Dye. Col.* **102** in the press
Mollon J D, Bowmaker J K, Dartnall H J A and Bird A C 1984a in *Colour Vision Deficiencies VII* (The Hague: W Junk) 303
Mollon J D, Bowmaker J K and Jacobs G H 1984b *Proc. R. Soc.* B **222** 373
Moran J J and Stonehill H I 1957 *J. Chem. Soc.* 765
Morley D I, Munn F and Billmeyer F W Jr 1975 *J. Soc. Dye. Col.* **91** 229
Morton J 1929 reprinted in *Fifty Years of Sundour* (Carlisle: Morton Sundour Fabrics Limited)
Morton T H 1976 *J. Soc. Dye. Col.* **92** 342
Müller G E 1930 *Zeit. Psychol. Ergänzungsber* **17**, **18**
Munz F W and Schwanzara S A 1967 *Vision Res.* **7** 111
Newhall S M, Nickerson D and Judd D B 1943 *J. Opt. Soc. Am.* **33** 385
Newton I 1730 *Opticks* reprinted 1952 (New York: Dover) p. 179
Nickerson D 1936 *Text. Res.* **6** 509
— 1950 *Am. Dyes. Rep.* **39** 541
— 1981 *Col. Res. Appl.* **6** 7
Nickerson D and Stultz K F 1944 *J. Opt. Soc. Am.* **34** 550
Norrish R G W and Porter G 1949 *Nature* **164** 658
Ohta N and Wyszecki G 1977 *Col. Res. Appl.* **2** 183
Ostwald W 1919 *Physik. Farbenlehre* (Leipzig: Hirsel)
Padgham C A and Rowe S C H 1973 *J. Col. Appl.* **II** (1) 18
Park J and Thompson T M 1982 *J. Soc. Dye. Col.* **98** 74
Park R H and Stearns E I 1944 *J. Opt. Soc. Am.* **34** 112
Parker C A and Hatchard C G 1961 *Trans. Farad. Soc.* **57** 1894
Pauli H 1976 *J. Opt. Soc. Am.* **66** 866
Pauli W E 1911 *Ann. Phys., Lpz.* **34** 739
Pellew C E 1919 *Col. Trade J.* **4** 97
Perkin W H 1856 *British Patent* 1984
Peters R H and Sumner H 1955 *J. Soc. Dye. Col.* **71** 130
Pineo O W 1940 *US Patents* 2 218 357; 2 194 910
Pointer M R 1981 *Col. Res. Appl.* **6** 108

Polyak S 1957 *The Vertebrate Visual System* (Chicago: Chicago University Press)
Porter G and Suppan P 1965 *Trans. Farad. Soc.* **61** 1664
Priest I G, Gibson K S and McNicholas H J 1920 *NBS Tech. Pap.* 167
Pritchard B S and Holmwood W A 1955 *J. Opt. Soc. Am.* **45** 690
Rabe P and Koch O 1957 *Melliand Textilber.* **38** 173
Ramsley A O 1967 *Col. Eng.* 5 (1) 20
Rattee I D 1965 *J. Soc. Dye. Col.* **81** 145
Rich D C and Billmeyer F W Jr 1979 *J. Coatings Technol.* **51** 650 p. 45
Robertson A R 1977 *Col. Res. Appl.* **2** 7
— 1978 *Col. Res. Appl.* **3** 149
— 1983 *Col. Res. Appl.* **8** 189
Robertson J M 1935 *J. Chem. Soc.* 615
Robinson G W and Frosch R P 1963 *J. Chem. Phys.* **38** 1187
Robinson R 1956 in *Proc. Perkin Centen. 1856–1956* (Research Triangle Park NC, USA: American Association of Textile Chemists and Colorists) p. 44
Rodrigues A B J and Besnoy R 1980 *Col. Res. Appl.* **5** 220
Rohner E 1965 in *Tag. Ber. Intern. Farbtagung Luzern* (Göttingen: Musterschmidt) p. 865
Rosenberg A and Shombert G 1960 *J. Am. Chem. Soc.* **82** 3527
Rushton W A H 1975 *Sci. Am.* **232** (3) 64
Sack R A and Seltzer S 1978 *Vision Res.* **18** 423
Saunderson J L 1942 *J. Opt. Soc. Am.* **32** 727
Schein S J, Marrocco R T and De Monasterio F M 1982 *J. Neurophysiol.* **47** 193
Schofield F and Goodyear J 1929 *Melliand Textilber.* **10** 867
Schofield F and Turner H A 1933 *J. Text. Inst.* **24** p. 131
Schorer M and Dörr F 1964 *Ber. Bunsenges.* **68** 478
Schouten J F and Ornstein L S 1939 *J. Opt. Soc. Am.* **29** 168
Schrödinger E 1920 *Ann. Phys. Chem.* **63** 481
— 1925 *Sitzber. Akad. Wiss. Abt. IIA* **134** 471
Schuster A 1905 *Astrophys. J.* **21** 1
SDC/AATCC 1971 *The Colour Index* 3rd edn (Bradford, England: Society of Dyers and Colourists; Research Triangle Park NC, USA; American Association of Textile Chemists and Colorists)
Sherman P D 1981 *Colour Vision in the 19th Century* (Bristol: Adam Hilger)
Shichi H 1983 *Biochemistry of Vision* (New York: Academic)
Shinozawa T, Uchida S, Martin E, Cafiso D, Hubbell W and Bitensky M 1980 *Proc. Natl. Acad. Sci.* **77** 1408
Shipp S D and Zeki S 1984 *J. Physiol.* **352** 23P
Simon F T 1961 *Die Farbe* **10** 225
— 1983 *AATCC–ISCC Symposium on Practical Applications of Color Control* p. 22
Simon F T and Goodwin W J 1957 *J. Opt. Soc. Am.* **47** 1050
Smart I C 1977 Private communication
Speranskaya N I 1959 *Opt. Spectrosc.* **7** 424
Stanziola R, Momiroff B and Hemmendinger H 1977 in *Color 77* (Bristol: Adam Hilger) p. 313
— 1979 *J. Col. App.* **4** 157
Stearns E I 1951 *Am. Dye. Rep.* **40** 563
Stiles W S 1978 *Mechanisms of Colour Vision* (London: Academic)

Stiles W S and Burch J M 1959 *Opt. Acta* **6** 1
Stiles W S and Wyszecki G 1968 *J. Opt. Soc. Am.* **58** 32
Strocka D 1971 Private communication from A Brockes 1975
Sullivan G A 1982 *Text. Chem. Col.* **14** 43
Svaetichin G 1953 *Acta Physiol. Scand.* **29** (suppl. 106) 565
— 1956 *Acta Physiol. Scand.* **39** (suppl. 134) 17
Taylor P F 1977 *Am. Soc. Qual. Cont.* **5** 67
Thornton W A 1974 *J. Ill. Eng. Soc. (USA)* **4** (Oct) 11
Thurner K and Walther V 1969 *Die Farbe* **18** 191
Thurner K and Würz A 1966 *Melliand Textilber.* **47** 1036
Tilleard D L 1964 *Paint Res. Stat. Bull.* **13** 161
Tomita T, Kaneko E, Murakami M and Pauter E L 1967 *Vision Res.* **7** 519
Troland L T 1922 *J. Opt. Soc. Am.* **6** 527
Tyminski P N, Klingbiel R T, Ott R A and O'Brien D F 1982 *Biochemistry* **21** 1197
Uyterhoeven W 1938 *Elektrische Gasentladungslampen* (Berlin: Springer)
Venkataraman K 1952 in *The Chemistry of Synthetic Dyes* vol. 2 (New York: Academic)
Vilter V 1949 *C.R. Soc. Biol.* **143** 830
Voke J 1980 *Colour Vision Testing* (London: Keeler)
Wald G 1945 *Harvey Lect. Ser.* **41** 117
— 1968 *Science* **162** 230
Walford J 1980 *Developments in Food Colours – I* (Barking, England: Applied Science)
Walls G L 1956 *J. Hist. Med.* **11** 66
— 1960 *Psychol. Bull.* **57** 29
Walls G L and Heath G G 1956 *J. Opt. Soc. Am.* **46** 640
Warburton P and Lund G V 1956 *J. Text. Inst.* **47** 305
Ward R A 1972 *J. Col. App.* **1** (6) 15
Wells C F 1956 *Nature* **177** 483
Werblin F S and Dowling J E 1969 *J. Neurophysiol.* **32** 339
White G S J 1960 *J. Soc. Dye. Col.* **76** 16
Wilkinson F, Seddon G M and Tickle K V 1968 *Ber. Bunseng. Phys. Chem.* **72** 315
Wilson M H and Brocklebank R W 1960 *J. Photogr. Sci.* **8** 141
— 1961 *Contemp. Phys.* **3** 91
Wilson M E and Cragg B G 1967 *J. Anat.* **101** 677
Witt O N 1876 *Berichte* **9** 522
Wollaston W H 1802 *Phil. Trans. R. Soc.* 365
Wong-Riley M T T 1976 *Brain Res.* **109** 595
Wright W D 1928 *Trans. Opt. Soc. Lond.* **30** 141
— 1929 *Trans. Opt. Soc. Lond.* **31** 201
— 1946 *Researches on Normal and Defective Colour Vision* (London: Kimpton)
— 1952 *J. Opt. Soc. Am.* **42** 509
— 1969 *The Measurement of Colour* 4th edn (London: Adam Hilger)
— 1970 *J. Col. Group* **15** 166
— 1981 in *Golden Jubilee of Colour in the CIE* (Bradford: The Society of Dyers and Colourists)
— 1982 *Col. Res. Appl.* **7** 12

Wyszecki G 1968 *J. Opt. Soc. Am.* **58** 290
— 1974 *J. Opt. Soc. Am.* **64** 896
Young T 1802a *Phil. Trans. R. Soc.* 12
— 1802b *Phil. Trans. R. Soc.* 387
— 1807 *A Course of Lectures on Natural Philosophy and the Mechanical Arts* republished 1845 (London: P Kelland)
Yukie M, Umibu J, Kido S, Niihara T and Iwai E 1979 *Neurosci. Lett.* **13** 544
Zeki S M 1971 *Brain Res.* **34** 19
— 1973 *Brain Res.* **53** 422
— 1980 *Nature* **284** 412
— 1984 *Proc. R. Inst.* **56** 231
Zweig A and Henderson W A Jr 1975 *J. Polym. Sci.* **13** 717

Author and Personal Names Index

Subject Index